THE ART AND CRAFT OF
GEOMETRIC ORIGAMI

THE ART AND CRAFT OF
GEOMETRIC ORIGAMI

MARK BOLITHO

PRINCETON ARCHITECTURAL PRESS · NEW YORK

Princeton Architectural Press
A McEvoy Group company
37 East 7th Street, New York, NY 10003
202 Warren Street, Hudson, NY 12534
www.papress.com

This book has been produced by
Jacqui Small LLP
74-77 White Lion Street
London, UK N1 9PF

For Jacqui Small
Commissioning and Project Editor: Joanna Copestick
Managing Editor: Emma Heyworth-Dunn
Senior designer and Art Director: Rachel Cross
Assistant Designer: Clare Thorpe
Photography: Brent Darby
Production: Maeve Healy

For Princeton Architectural Press
Project Editor: Barbara Darko

ISBN 978-1-61689-634-8

20 19 18 17 4 3 2 1

Printed in China

Scaling and Sizes

Each project is accompanied by a scaling
diagram that shows the size of the final model compared to the starting sheet of paper.
The diagram is based on a square sheet
with dimensions of 7 × 7 in. (18 × 18 cm), or
an equivalent rectangle. However, larger or
smaller models can be made. The size of
the final model can be scaled up or down by
comparing the dimensions of the paper used
to the sheet used in the scaling diagram.

Complexity Ratings

The projects in this book have been given a
rating based on their complexity, which ap-
pears beneath the title of each project.

Easy ✳

Intermediate ✳✳

More Challenging ✳✳✳

CONTENTS

MINDFUL ORIGAMI

Welcome to the world of origami—the art of paper folding. At its heart it is the transformation of a sheet of paper into a finished model. However, it's not only a matter of creating a finished model, but also a journey of paper folding that involves creativity and contemplation along the way to produce your finished piece of work.

The word *origami* comes from the Japanese word for "paper folding." In the East the craft developed based on standard forms and traditional designs, and it now has many enthusiasts around the world.

The internet has enabled the sharing of ideas and led to a collective enthusiasm for developing more beautiful and complex designs. In the chase for complexity, however, some of the beauty of the craft has been overlooked. This is a discrepancy I hope to redress with this book by presenting mindful finished works in appropriate colors and compositions.

The paper-folding process can be a contemplative one; over time a plain sheet of paper is transformed into something wonderful. The satisfaction of origami comes not only from creating interesting designs, but also from following the folding journey and seeing your model evolve at your fingertips. Origami offers a perfect way to explore your mindful creativity in the colors and paper choices you use. In addition, you can explore and consider the nature of paper itself and the final composition of groups of complementary models.

The projects in this collection have been selected based on the aesthetic quality of the final model and the folding processes. They are explained with step-by-step diagrams that show the sequences needed to produce the final design.

At the start of the collection I have included instructions for the Fortune Teller. This is an opportunity to gain familiarity with the diagrams and symbols used to explain the folding sequences. Some models are more complex than others and we have given a rating to each project as a guide (see page 4).

If you are new to origami, try starting with the easier models and working up to more complex projects. I hope you enjoy folding these projects as much as I enjoyed designing them.

GETTING STARTED

Here are the basic folding techniques and symbols you will need to complete the projects in the book.

FOLDING IN HALF

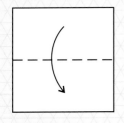 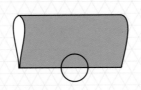 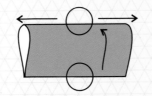 

1. The step indicates that the paper should be folded in half.

2. First of all, line up the opposite sides of the paper and hold the edges together.

3. When the two layers are aligned, pinch the middle to hold them together. Then make the crease.

4. The paper is accurately folded in half.

ARROWS AND FOLDS

Paper-folding instructions explain the folding process with a series of steps leading to a finished model. Each step explains one or two folds in the process. Steps should be followed in order, and when a step is completed it should resemble the image shown in the next step.

The transition from one step to the next is shown by a series of lines and arrows indicating where folds should be made. The lines show where to fold and the arrows show how the paper should be moved to make folds.

Folds are described as either Mountain Folds or Valley Folds.

These names refer to how the surface will look after the fold has been completed. A Mountain Fold will fold toward the observer, forming a mountain shape, while a Valley Fold will fold away, forming a *V* or valley shape. They are represented by differing dotted-line symbols.

ARROWS

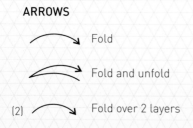

Fold

Fold and unfold

(2) Fold over 2 layers

FOLDS

	Description	In Progress	Completed

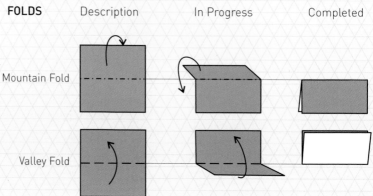

Mountain Fold

Valley Fold

ORIGAMI SYMBOLS

Various symbols are used to explain the folding process, such as turning the model over, rotating the model, or repeating a step. The symbols on the right are the ones used in this book.

Follow the instructions in numerical order. After completing the folds in any step, look ahead; the model you have should look like the diagram in the next step. If not, undo the folds and try again. Each step is self-contained, with additional information in the caption below each diagram.

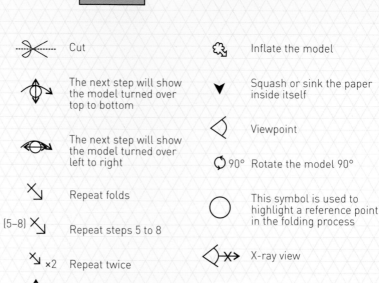

Cut

The next step will show the model turned over top to bottom

The next step will show the model turned over left to right

Repeat folds

(5–8) Repeat steps 5 to 8

×2 Repeat twice

(8) Unfold to step 8

Inflate the model

Squash or sink the paper inside itself

Viewpoint

○ 90° Rotate the model 90°

This symbol is used to highlight a reference point in the folding process

X-ray view

DIAGRAMS

The diagrams are shown in two colors, with the colored side being the front and the white side being the reverse. This should make the step-by-step instructions easier to follow.

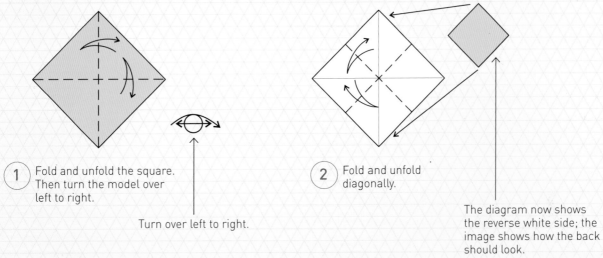

(1) Fold and unfold the square. Then turn the model over left to right.

Turn over left to right.

(2) Fold and unfold diagonally.

The diagram now shows the reverse white side; the image shows how the back should look.

THE FOLLOWING IS A LONGER SEQUENCE TO MAKE A PRELIMINARY BASE.

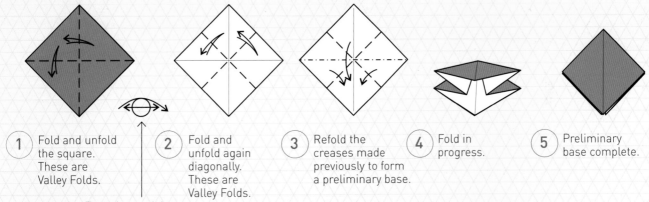

(1) Fold and unfold the square. These are Valley Folds.

Turn the model over.

(2) Fold and unfold again diagonally. These are Valley Folds.

(3) Refold the creases made previously to form a preliminary base.

(4) Fold in progress.

(5) Preliminary base complete.

Origami instructions take you through a step-by-step process from start to finish. Symbols are included to explain the transition from one step to the next. Each step shows how the folded project should look and shows the folds that should be applied to progress to the next step.

When approaching a step, look ahead to the next diagram to see how the model should look when the fold has been applied. Then check and make sure that the paper model you are making resembles the step diagram. Look out for reference points to compare your model

with the instructions to make sure you remain on the right track.

If your model doesn't resemble the step you are on, unfold the last step and work back until it does.

Fortune Teller

*

The classic fortune teller is an introductory model included to enable familiarization with the diagrams and symbols used. The asterisk (*) indicates that this is an easy model.

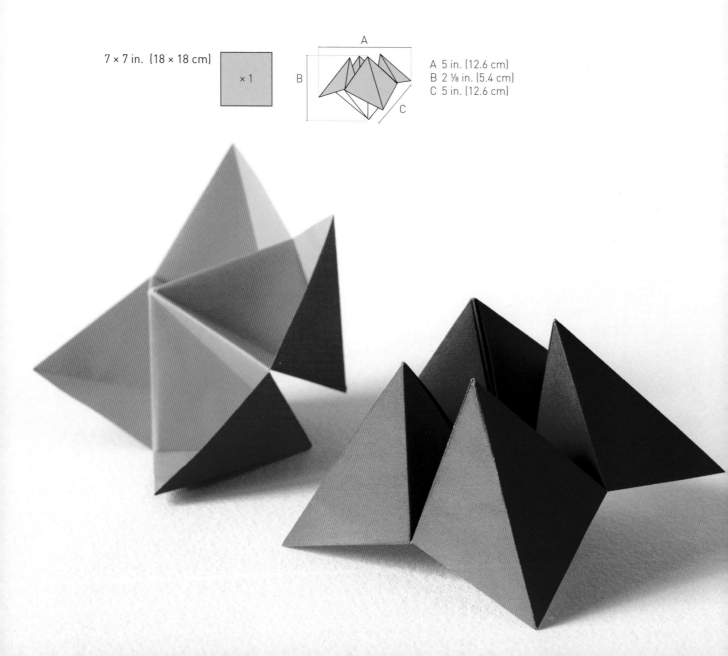

7 × 7 in. (18 × 18 cm)   ×1

A 5 in. (12.6 cm)
B 2 ⅛ in. (5.4 cm)
C 5 in. (12.6 cm)

START WITH A SQUARE, COLORED SIDE UP.

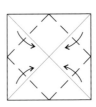

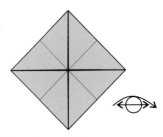

1 Fold and unfold the square in half diagonally. Then turn the model over left to right.

2 Fold the corners into the center.

3 Turn the model over left to right.

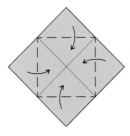

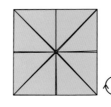

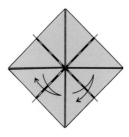

4 Fold the corners into the center.

5 Rotate the model 45°.

6 Fold and unfold along the diagonals as shown.

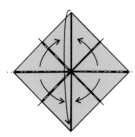

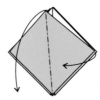

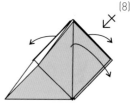

7 Fold the top section down. At the same time, fold in the corners and make a preliminary base shape.

8 Fold one side over and fold out the top layer.

9 Repeat step 8 on the other three sides.

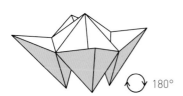

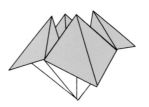

10 Rotate the model 180° top to bottom.

11 Complete.

ONE-PIECE
PROJECTS

Tetrahedron

The tetrahedron is formed by dividing the paper into three, then folding
four equilateral triangles to become the faces of the assembled tetrahedron.
The final look is subtle but holds firm. The tetrahedron is the first of the
platonic solids and in early writings is associated with fire.

7 × 7 in. (18 × 18 cm)   × 1

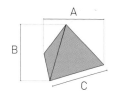

A   2 ¾ in. (7 cm)
B   2 ¼ in. (5.7 cm)
C   2 ¾ in. (7 cm)

START WITH A SQUARE, COLORED SIDE UP.

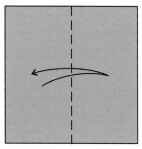

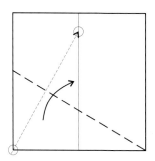

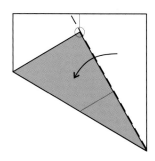

(1) Fold and unfold the square in half lengthwise. Then turn the model over left to right.

(2) Fold the bottom left corner up to touch the center crease.

(3) Fold the opposite side over.

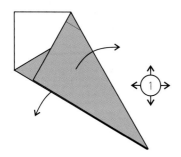

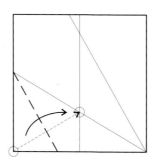

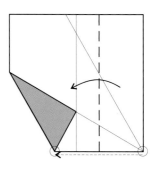

4. Unfold back into a square.

5. Fold the bottom left corner in so that the edge of the section is aligned with the crease.

6. Fold the right edge over. The bottom corner should touch the folded corner made previously.

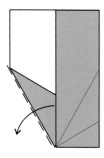

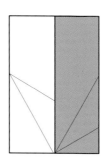

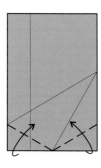

7. Fold the left corner back out.

8. Turn the model over left to right.

9. Fold the bottom corners up. The edges should touch the adjacent diagonal creases.

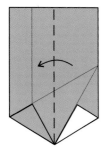

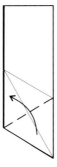

10. Fold one side over to fold the model in half.

11. Fold up the bottom corner.

12. Fold down the top section. The outer edge should touch the edge of the bottom section.

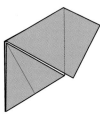

13 Turn the model over left to right.

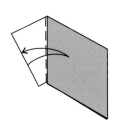

14 Fold the top section in and out again along the adjacent edge of the paper.

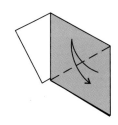

15 Fold and unfold between the outer corners of the diamond shape.

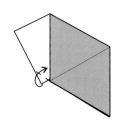

16 Fold the outer corner in.

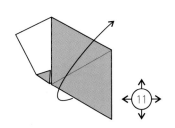

17 Unfold the previous folds to step 11, but leave the corner folded over in step 16 folded.

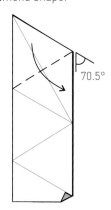

18 Fold the top corner in at an angle.

19 Fold the bottom section up so that the edges of the bottom and top sections touch.

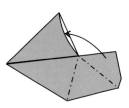

20 Fold the paper around and tuck the corner into a pocket formed by the layers of paper.

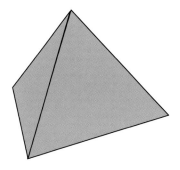

21 Complete.

Cube

*

The cube is the second of the platonic solids, a shape made from six square faces. The model is folded by developing the geometry of the square and making smaller squares within the larger one. As a platonic solid, the cube is associated with the earth.

7 × 7 in. (18 × 18 cm)

× 1

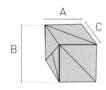

A
B
C

A   1 ¾ in. (4.5 cm)
B   1 ¾ in. (4.5 cm)
C   1 ¾ in. (4.5 cm)

START WITH A SQUARE, COLORED SIDE UP.

(1) Fold and unfold the square in half lengthwise. Then turn the model over left to right.

(2) Fold the edges into the center crease and unfold.

(3) Repeat the process on the other axes to divide the paper into 16 sections.

(4) Fold and unfold the square diagonally.

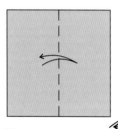

(5) Fold and unfold diagonally. The line styles show Valley or Mountain Folds.

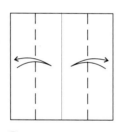

(6) Fold the sides in so they are perpendicular to each other.

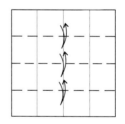

(7) Fold the corner up to be flush with the side of the model.

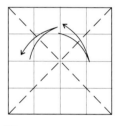

(8) Fold edge (a) down first, then fold (b) over it so they are both perpendicular to the bottom section.

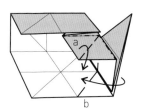

(9) Fold one side in, at the same time folding in the corner at (a) and making a diagonal fold at (b).

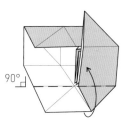

(10) Fold the edge up.

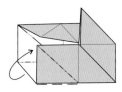

(11) Reverse fold the bottom section into the model along two diagonal folds.

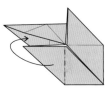

(12) Fold the triangular corner in to touch the side of the cube.

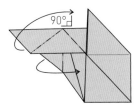

(13) Fold the top section upward, then fold it flat against the side of the cube.

(14) Fold the top section up flat against the top of the cube. Tuck one triangular point into the other.

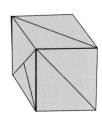

(15) Complete.

Hexahedron

This model starts from the classic fortune teller (see page 10) and then uses the spaces in the model to make a solid shape. The final step is particularly pleasing, as it ends with a corner of the model being tucked into a pocket created in the edge of the regular solid.

7 × 7 in. (18 × 18 cm)   × 1

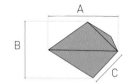

A   2 5/8 in. (6.6 cm)
B   2 1/8 in. (5.4 cm)
C   2 3/8 in. (6 cm)

START WITH A SQUARE, COLORED SIDE DOWN.

(1) Fold and unfold the square in half lengthwise and diagonally along all axes.

(2) Fold the corners into the center.

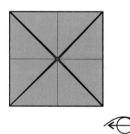

(3) Turn the model over left to right.

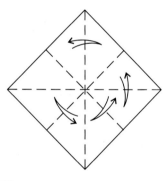

(4) Fold the corners into the middle again.

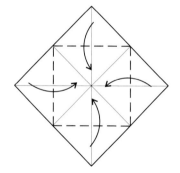

(5) Rotate the model 45°.

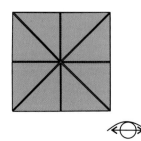

(6) Turn the model over left to right.

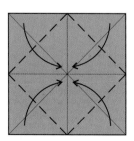
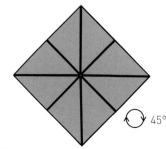

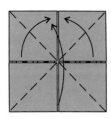

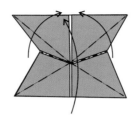

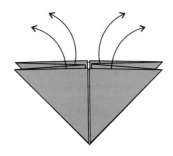

7 Fold the bottom section up and the corners in to form a waterbomb base.

8 Fold in progress.

9 Reverse the trapped paper out from the inside of the folded corners.

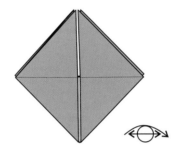

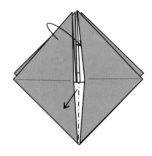

10 Fold the model in half and then unfold.

11 Turn the model over left to right.

12 Fold out the inner edge. This will open out the adjacent paper into a 3-D shape.

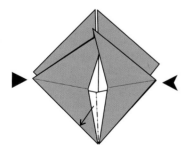

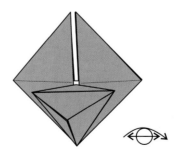

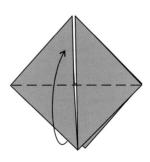

13 Push the sides together and tuck one side into the other.

14 The assembled sections will form a hexahedron. Turn the model over left to right.

15 Fold the corner up.

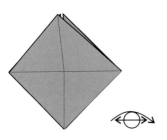

16 Turn the model over left to right.

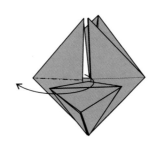

17 Pull out the adjacent edge and open it up.

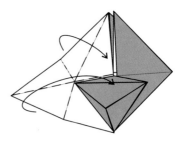

18 Wrap the outer paper around the hexahedron shape.

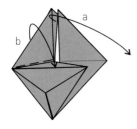

19 Open out the paper by pulling out the corner (a) and continue wrapping the hexahedron shape by folding over at (b).

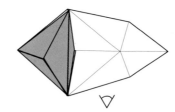

20 Rotate the model slightly and look at its side.

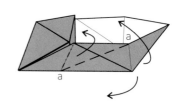

21 Wrap the outer paper around by making a crease along the (a–a) edge.

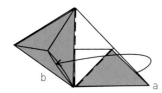

22 Fold the pinched corner up and tuck the point (a) into the pocket (b).

23 Continue tucking the point into the pocket.

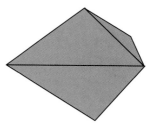

24 Complete.

Octahedron One

*

The octahedron is an eight-sided regular solid constructed from equilateral triangles. It wraps together neatly, with the corners folding into pockets to complete the project.

7 × 7 in. (18 × 18 cm) × 1

A 2 in. (5.2 cm)
B 3 in. (7.6 cm)
C 2 in. (5.2 cm)

START WITH A SQUARE, COLORED SIDE UP.

1. Fold and unfold the square in half lengthwise. Then turn the paper over left to right.

2. Fold and unfold the square diagonally along the middle.

3. Fold and unfold the outer edges into the middle crease.

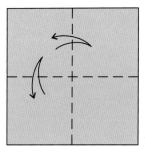

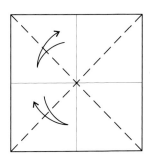

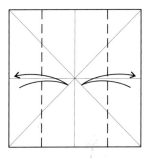

4. Fold and unfold the top and bottom edges into the middle crease.

5. Fold the paper in half so that the top edge touches the bottom edge.

6. Fold the corners in so the outer corner touch the vertical creases.

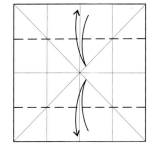

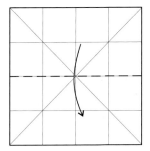

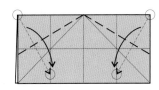

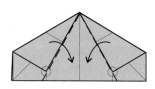

(7) Fold the sides in again along the folded edges.

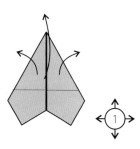

(8) Unfold the paper back into a square.

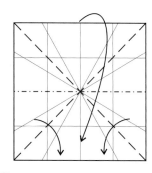

(9) Fold the top section down and the sides in to make a waterbomb base.

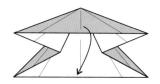

(10) Fold in progress.

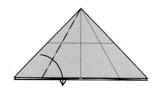

(11) Fold one side in along the crease made previously.

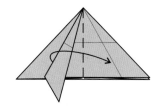

(12) Fold one side over to the right.

(13) Fold the corner up into the pocket formed by the layers of the top section.

(11–13)

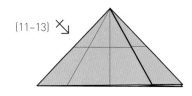

(14) Repeat steps 11 to 13 on the other side of the model.

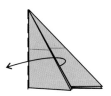

(15) Fold the corner back again.

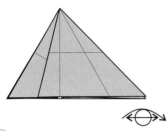

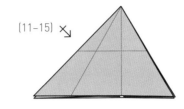

(11–15)

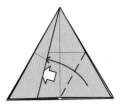

16 Turn the model over left to right.

17 Repeat steps 11 to 15.

18 Fold the corner in and tuck it underneath the folded edge on the opposite side.

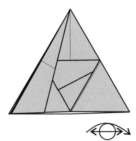

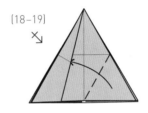

(18–19)

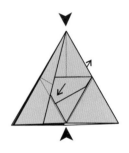

19 Fold and unfold the corner along the adjacent folded edge.

20 Turn the model over left to right.

21 Fold the corner in and repeat steps 18 to 19.

22 Open out the model and pinch the two remaining corners together.

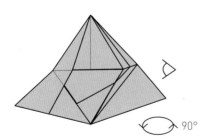

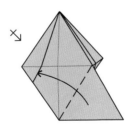

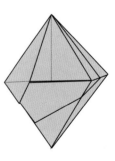

23 Rotate the model to look at the other sides.

24 Fold the corners of the front and rear sections into the pockets opposite.

25 Complete.

Octahedron Two

✳✳

This octahedron is formed by dividing the starting square into thirds and using the geometry formed by the creases. The inverted point adds rigidity to the model.

7 × 7 in. (18 × 18 cm)

 × 1

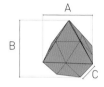

A   2 ⅛ in. (5.3 cm)
B   2 ¼ in. (5.8 cm)
C   2 ⅛ in. (5.3 cm)

START WITH A SQUARE, COLORED SIDE UP.

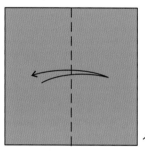

(1) Fold and unfold the square in half lengthwise. Then turn the model over left to right.

(2) Fold the outer sides into the center crease and unfold.

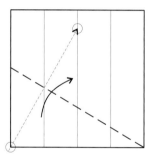

(3) Fold the bottom section up diagonally from the bottom right corner. The left corner will touch the center crease.

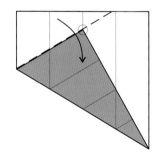

(4) Fold the top left corner down over the folded edge.

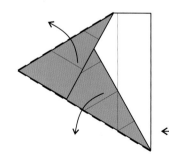

(5) Unfold back into a square.

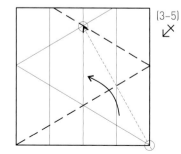

(6) Repeat steps 3 to 5 on the other side.

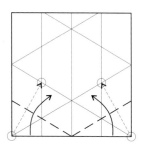

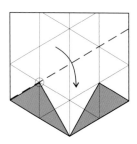

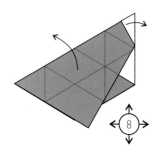

(7) Fold the bottom corners up to touch the adjacent creases.

(8) Fold the top section over along the adjacent folded edge.

(9) Fold the top right corner over.

(10) Unfold the last two steps (to step 8).

(8–10)

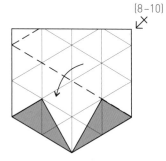

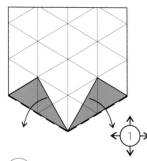

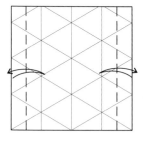

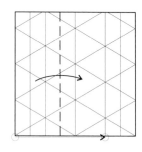

(11) Repeat steps 8 to 10 on the other side.

(12) Unfold the bottom corners. (Back to a square, step 1).

(13) Fold the outer edges into the adjacent creases, then unfold.

(14) Fold the left edge over between the two creases.

(14–15)

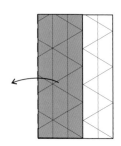

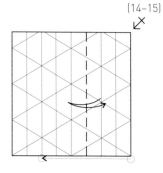

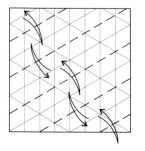

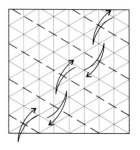

(15) Fold it back.

(16) Fold and unfold the right edge between the two creases.

(17) Fold and unfold between the diagonal creases made previously.

(18) Repeat the process on the opposite side.

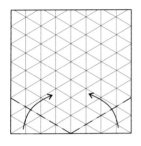

19 Fold the bottom corners back up.

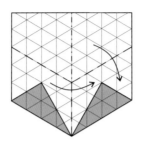

20 Fold the paper in half and at the same time reverse fold the top section to the right.

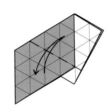

21 Fold the top section over and back again.

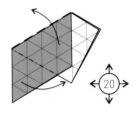

22 Unfold to step 20.

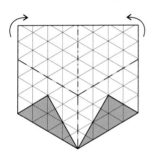

23 Fold the sides together and fold the top section up along the creases made previously.

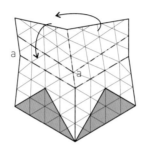

24 Fold the top section in along (a–a) to make an inverted pyramid in the middle of the model.

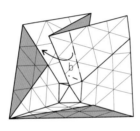

25 This shows the inverted pyramid in the center. Continue folding the section (b) around it.

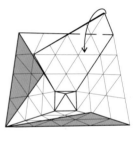

26 Fold the outer edge over and into the model.

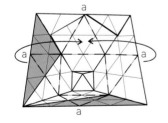

27 Fold the sides in, bringing all of the points (a) together.

28 Fold one of the equilateral triangles into the other. One should tuck fully inside the other.

29 Fold the triangles on the opposite side into each other.

30 Complete.

Sunken Pyramid

*

This project starts with the geometry of a square. The sides are formed by three-quarters of the square folding together. The inverted point is part of the look of the model on the outside. It also works inside the model, applying an internal tension that adds rigidity.

7 × 7 in. (18 × 18 cm)   ×1

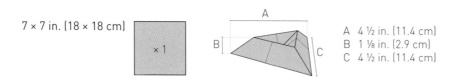

A   4 ½ in. (11.4 cm)
B   1 ⅛ in. (2.9 cm)
C   4 ½ in. (11.4 cm)

START WITH A SQUARE, COLORED SIDE UP.

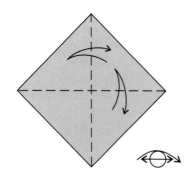

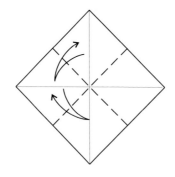

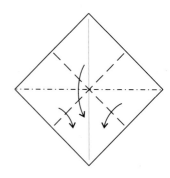

1 Fold and unfold the square diagonally along both axes. Then turn the paper over left to right.

2 Fold and unfold along the center of the paper on both axes.

3 Fold the top corner down and fold the outer corners in, making a preliminary base.

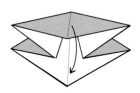

4 Fold in progress.

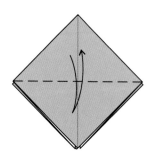

5 Fold the bottom corner up and down again, along the center.

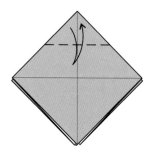

6 Fold and unfold the top corner into the center crease.

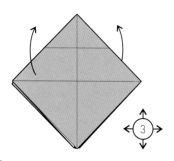

7 Unfold back into a square (to step 3).

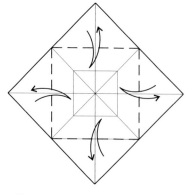

8 Fold the corners into the center and unfold.

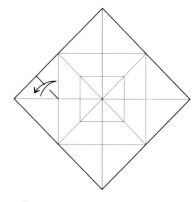

9 Fold and unfold the outer left edge diagonally.

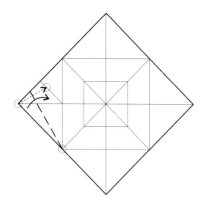

10 Fold the left corner into the crease made previously.

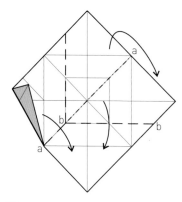

11 Fold the model in half along (a–a). At the same time fold down the line (b–b) to remake the model along the creases made previously.

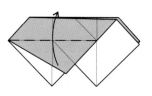

12 Fold up the bottom corner of the top layer.

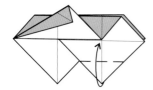

(13) Fold up the corner of the bottom section.

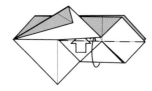

(14) Fold the bottom edge up and into the model.

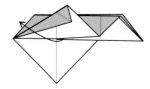

(15) Open out the model by folding the top edges to the left.

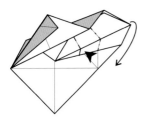

(16) Having opened out the model, push the paper inward where indicated to form the third edge of a triangle.

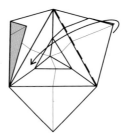

(17) This view shows the underside. Fold the outer corner over.

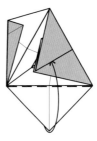

(18) Fold the second side up and tuck it under the folded edge.

(19) Fold the final side over and tuck the folded flap (a) beneath the folded edge (b).

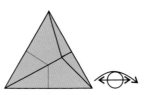

(20) The edges should hold together as a result of the tension from the folded section beneath. Turn the model over.

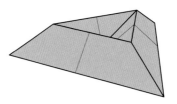

(21) Complete.

Diamond

*

The diamond is constructed from radial folds emanating from the center of a square.
This forms the narrower end, with the edges of the square folding
in on themselves to form the face.

7 × 7 in. (18 × 18 cm)

× 1

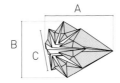

A   2 7/8 in. (7.2 cm)
B   1 7/8 in. (4.7 cm)
C   1 7/8 in. (4.7 cm)

START WITH A SQUARE, COLORED SIDE UP.

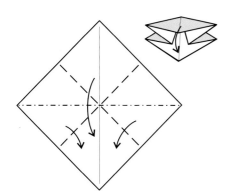

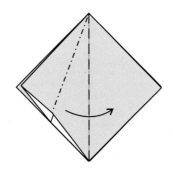

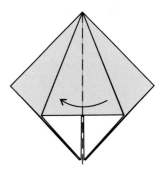

1. Fold the top corner down to touch the opposite corner. At the same time fold in the sides to make a preliminary base for the model.

2. Fold the left corner up, separate the layers, and squash flat.

3. Fold the right side back to the left.

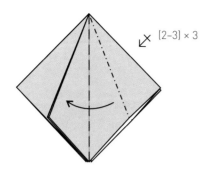

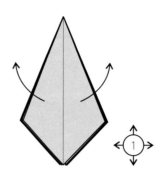

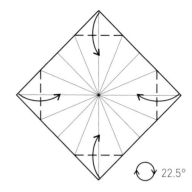

4 Repeat the squash process on the other three corners.

5 Open out the paper back into a square (step 1).

6 Fold the four corners in between the creases made previously. This will make an octagon. Then rotate the model.

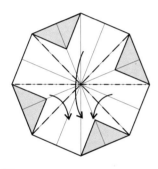

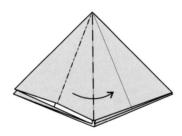

7 Fold the top section over the bottom half, and at the same time fold in the sides.

8 Fold the corner up, separate the layers, and squash flat.

9 Repeat the squashing process on the other three corners.

10 Fold the edges into the center crease.

11 Fold and unfold the top section over the folded edges.

12 Fold the corners back out again.

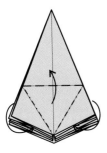

13 Fold the top layer up along the crease made previously. This will cause the outer edges to fold in.

14 Fold the tip of the corner down to touch the crease.

15 Fold the section over again.

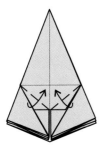

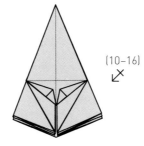

16 Fold the corners in on both sides.

(10–16)

17 Repeat steps 10 to 16 on the other faces.

18 Open the model and even out the layers.

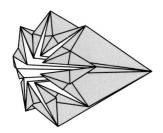

19 Complete.

Geo Ball

**

The geo ball works through the buildup of a curved geometric texture that allows the model to wrap around on itself. The ball is completed by tucking one end into the other. Once the ball is assembled it can be squashed gently by applying pressure at either end.

3 ½ × 10 ⅝ in. (9 × 27 cm)   ×1

A   5 ⅓ in. (13.5 cm)
B   4 ½ in. (11.6 cm)
C   5 ⅓ in. (13.5 cm)

START WITH A 3 × 1 RECTANGLE, COLORED SIDE DOWN.

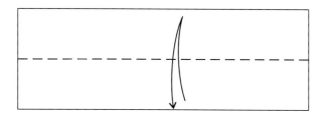

1. Fold and unfold the rectangle in half lengthwise.

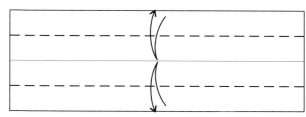

2. Fold the edges into the center crease and unfold.

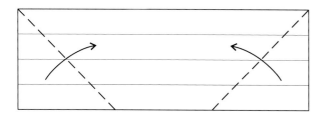

3. Fold the bottom corners in so that the outer edges touch the top edge of the rectangle.

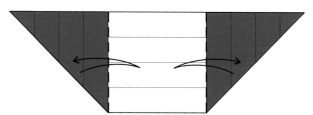

4. Fold the sides in along the edges of the folded triangles, then unfold.

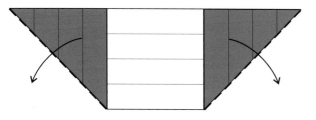

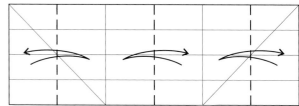

(5) Unfold the triangles.

(6) Fold and unfold between the creases made previously.

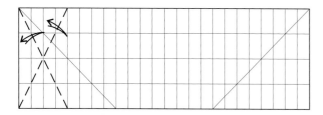

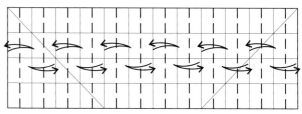

(7) Fold and unfold between the creases made previously.

(8) Fold and unfold between the creases made previously.

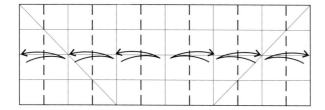

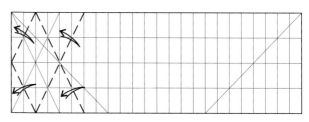

(9) Fold and unfold diagonally.

(10) Continue folding the diagonals across the rectangle shape.

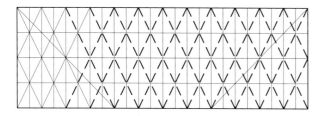

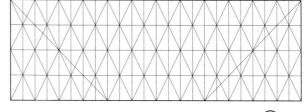

(11) Continue the diagonal fold across the rectangle.

90°

(12) Rotate the model 90°.

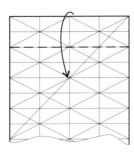

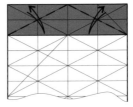

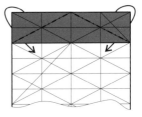

13 The following steps show a detail of the top section. Fold the edge down.

14 Fold and unfold the outer corners of the paper.

15 Reverse fold the corners inside the folded edge.

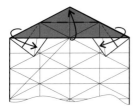

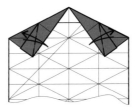

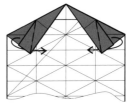

16 Fold the top layer up to touch the top corner, causing the edges to fold in.

17 Fold and unfold the outer corners of the paper.

18 Reverse fold the corners inside along the creases made previously.

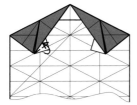

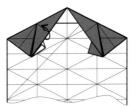

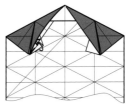

19 Fold the inside left corner in.

20 Open up the edge above the folded corner.

21 Fold the folded corner in between the layers of the adjacent edge.

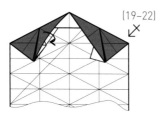

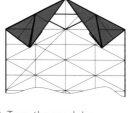

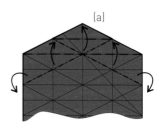

(a)

22 Fold the edge back and then repeat steps 19 to 22 on the other side.

23 Turn the model over left to right.

24 Fold top edge (a) up and fold up the paper below along the creases made previously.

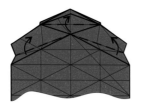

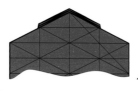

(19–22)   (19–22)

25 This shows the previous fold in progress.

26 Turn the model over left to right.

27 Repeat steps 19 to 22 on both sides, folding and locking the corners.

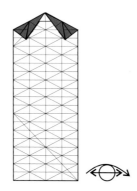

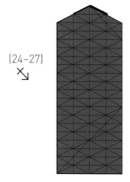

(24–27)

28 Turn the model over left to right.

29 Fold the rest of the paper up, repeating steps 24 to 27 along the length of the model.

30 Curve the model to align the front and rear faces.

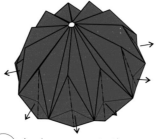

31 Tuck section (a) inside the other (b).

32 Apply pressure to the top and bottom, and pop out the corners around the middle to add more rigidity.

33 Complete.

GEO BALL ASSEMBLY

31a

Align the two opposite ends, squeezing the model until both ends are roughly the same size.

31b

Insert one end into the other. Match the ridges of one end with the equivalent of the other.

32

When the two ends are aligned, apply pressure at both ends to match and fit them together to complete the ball.

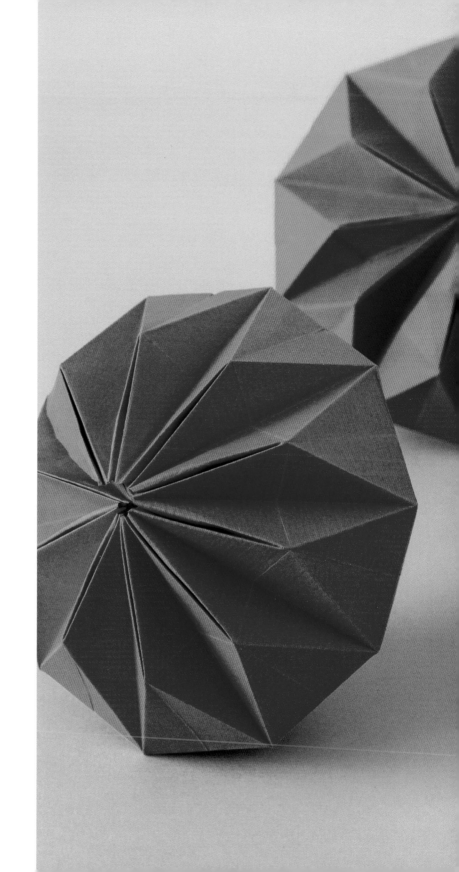

Icosahedron

***

The icosahedron works through rotational symmetry. The two sides of the model fold as mirror images of each other. The final assembly involves one end being tucked into the other and can be a bit challenging to achieve, but it does fit together neatly.

3 ½ × 7 in. (9 × 18 cm)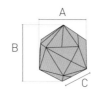

A   1 ⅞ in. (4.7 cm)
B   1 ⅞ in. (4.7 cm)
C   1 ⅞ in. (4.7 cm)

START WITH A 2 × 1 RECTANGLE, COLORED SIDE DOWN

(1) Fold and unfold the rectangle in half lengthwise.

(2) Fold and unfold the edges into the center crease.

(3) Fold the bottom corner up to touch the center crease. The fold should start from the opposite top corner.

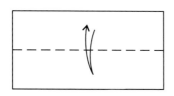

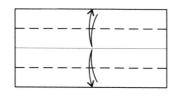

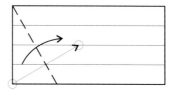

(4) Fold the top section over along the edge.

(5) Fold the section over the edge again.

(6) Fold the outer corner in over the edge of the triangle.

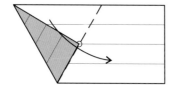

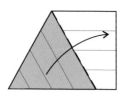

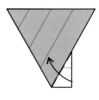

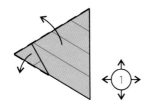

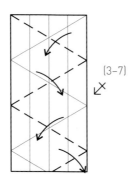

(3–7)

7 Unfold back to the starting rectangle.

8 Repeat steps 3 to 7 on the other side. This will mirror the creases made previously.

9 Fold and unfold between the parallel diagonal creases.

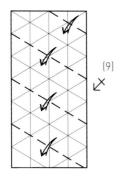

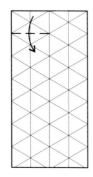

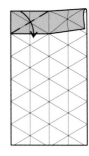

(9)

10 Fold and unfold between the parallel creases. This will mirror the previous step.

11 Fold the top edge down at the first intersection of the diagonal creases. This fold should only be on the left side of the paper.

12 Make a diagonal fold in the folded section along the creases made previously.

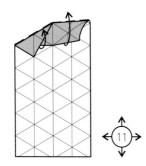

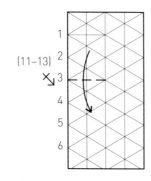

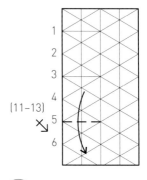

(11–13)

(11–13)

13 Unfold steps 11 and 12.

14 Repeat steps 11 to 13 at the third intersection of the diagonal creases.

15 Repeat steps 11 to 13 at the fifth intersection of the diagonal creases.

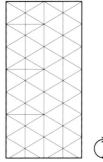

(16) Rotate the model 180°.

(17) Repeat steps 11 to 13 on the first, third, and fifth crease intersections. Note these are not aligned with the creases made on the opposite side.

(18) Fold the bottom left corner up.

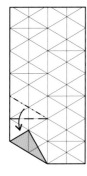

(19) Fold the paper above the corner over and back again along the creases made previously in the paper.

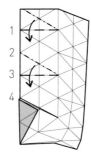

(20) Repeat the folding process in step 19 on the creases at the first and third intersections.

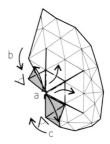

(21) Hold all of the folded sections together and fold the inner section (a) in. This will bring the sides (b) and (c) together.

(22) Fold the edge over and into the model.

(23) Fold the corner (a) into the space (b).

(24) This shows a view of the fold from the top. The folded corner (a) is folding over (b) to form an inverted pyramid made of three equilateral triangles.

25 This shows one edge wrapped together. Next, unfold back to the starting rectangle.

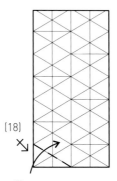

26 Fold the bottom left corner in, repeating step 18.

(18)

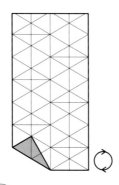

27 Rotate the model 180°.

180°

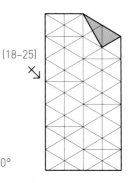

28 Repeat steps 18 to 25 to fold and roll up the edge of the paper.

(18–25)

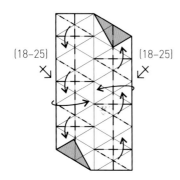

29 Now refold both edges of the rectangle at the same time, using the creases made previously.

(18–25)   (18–25)

30 The model is now a bit twisted. Push the two sides together to tuck one side into the other (see below).

31 The two sides should slide together with the top section sliding into the bottom section.

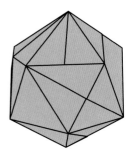

32 The icosahedron should lock together. Rotate it slightly and stand it up.

FINAL ASSEMBLY OF THE ICOSAHEDRON

30a Hold the folded sides together and line them up.

30b Open one slide slightly and bring the two sides together.

31a Insert one side into the other. The triangles of each side should align.

31b Check that you have completed step 22 with no obstructions, then slide to align.

32 Complete.

Pyramid

*

The pyramid is made by folding the corners of a square into each other to form the shape. The folding sequence builds an external skin and the final model is made three dimensional by inflating the model.

7 × 7 in. (18 × 18 cm)

×1

A 2 ⅔ in. (6.7 cm)
B 2 ⅛ in. (5.5 cm)
C 2 ⅔ in. (6.7 cm)

START WITH A SQUARE, COLORED SIDE UP.

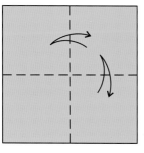

1. Fold and unfold the square in half lengthwise. Then turn the paper over left to right.

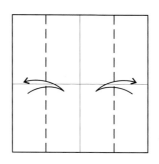

2. Fold and unfold the sides of the square into the middle crease.

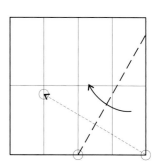

3. Fold the bottom right corner of the square into the crease. The fold will start from the middle crease.

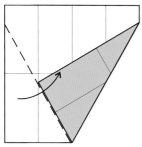

4. Fold the opposite side over the folded edge.

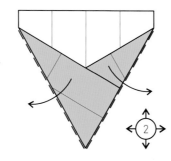

5. Unfold back into a square.

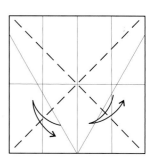

6. Fold and unfold the square in half diagonally.

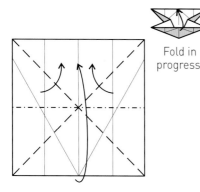

Fold in progress.

7. Fold the model together to make a waterbomb base.

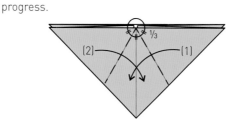

8. Fold the outer corners in along the creases made previously.

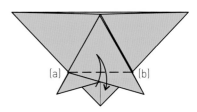

9. Fold and unfold the top layer along (a—b).

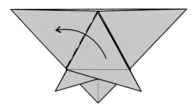

10 Fold one side up.

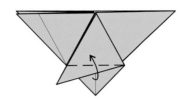

11 Fold the edge of the bottom section up along the crease made previously.

12 Fold the bottom corner up over the adjacent folded edge, then unfold.

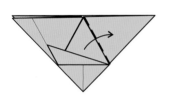

13 Fold the top layer up.

14 Fold the corners over and into the model between the layers.

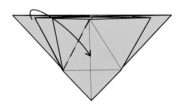

15 Fold one side over.

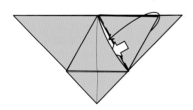

16 Fold the corner over and tuck it into the adjacent corner.

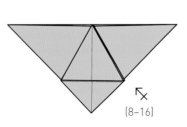

17 Repeat steps 8 to 16 on the reverse.

(8–16)

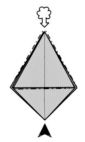

18 Inflate the model by pushing the base in and opening the sides.

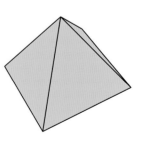

19 Complete.

TWISTS AND TURNS

Twister

The twister is an abstract shape constructed by applying a series of regular diagonal folds. These folds cause the model to curve. When pressure is applied to both sides, the model pops up and one side slides into the other. The final folds lock the model in place.

5 ⅓ × 7 in. (13.5 × 18 cm)

× 1

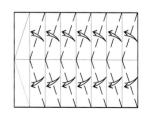

A 2 in. (5 cm)
B 3 ½ in. (9 cm)
C 2 in. (5 cm)

START WITH A 3 × 4 RECTANGLE, COLORED SIDE DOWN.

(1) Fold and unfold the paper in half lengthwise.

(2) Fold and unfold the center crease.

(3) Fold and unfold the edges to the center crease.

(4) Fold and unfold between the creases made previously.

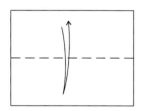
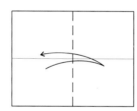
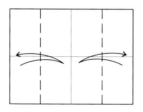
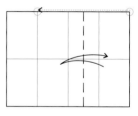

(5) Fold and unfold between the outer edge and the adjacent crease.

(4–5)

(6) Repeat steps 4 to 5 on the other side.

(7) Fold and unfold diagonally across the space defined by the creases.

(8) Repeat this process along the rest of the paper.

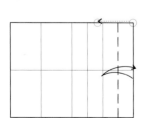
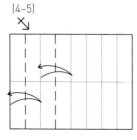
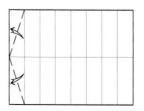

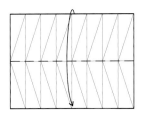

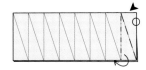

9 Fold the paper in half lengthwise, so the top edge touches the bottom edge.

10 Hold the outer edge of the model and fold the corner into the model along the creases made previously.

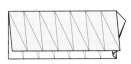

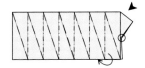

11 This shows the fold in progress. Note that the crimp fold takes the corner inside the model.

12 Repeat this process on the adjacent creases.

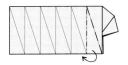

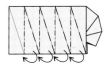

13 Continue this process and start to roll the paper around as you go.

14 Complete the crimping process for the rest of the model.

15 The crimping process is complete. Now fold over the corner of the last section.

16 Fold and unfold the corner aligned with the folded edge beneath.

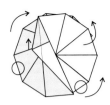 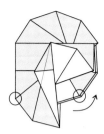 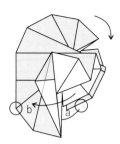 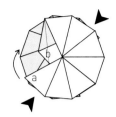

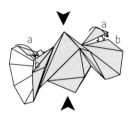   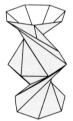

(17) Now open up the model by holding the sides at the *O*'s and opening out. The aim is to pull out trapped corners and slide them outside.

(18) Continue opening the model.

(19) Now fold the model back together, but fold the corners (a) over the outer layers (b).

(20) Push the sides of the model together. This will extend the model and push the edges (a) and (b) together. The next step will show a side view of the model.

(21) Continue pushing the body together and align the outer edges by sliding the outer edge (a) over the adjacent edge (b) at both ends.

(22) Rotate the model so it stands on one end.

(23) With the ends aligned, fold over the crease made in step 16 to link two layers together. Repeat this on the bottom section.

(24) Complete.

OPENING UP THE MODEL

20

This photo shows a side view of the two sides being pushed together.

21

Continue pushing the sides and extending the model.

22

The two sides should fit together in preparation to lock them together by folding the corners over in step 23.

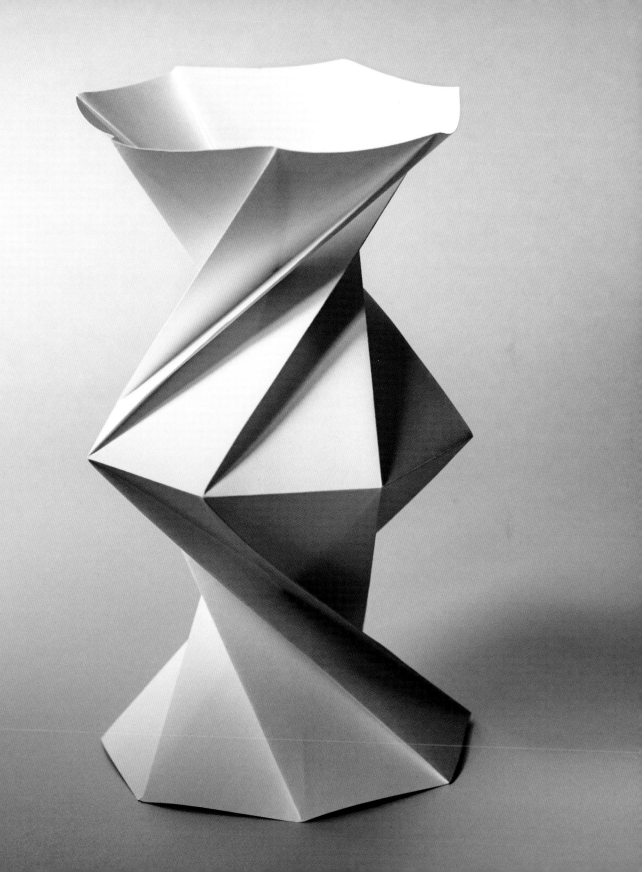

DNA Wheel

**

The DNA wheel fold builds a spiral similar to a natural DNA structure. Within the folding process there is a fluidity to the model. The opposite sides of the square twist, turning one side over until the ends of the wheel tuck into each other, locking them together.

7 × 7 in. (18 × 18 cm)

x 1	

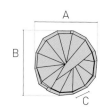

A   3 ½ in. (9 cm)
B   3 ½ in. (9 cm)
C   ¾ in. (2 cm)

START WITH A SQUARE, COLORED SIDE UP.

(1) Fold and unfold the square in half lengthwise.

(2) Fold the edges into the center crease and unfold.

(3) Fold and unfold between the creases to divide the paper into eight sections.

(4) Fold and unfold the side edges in to touch the center crease.

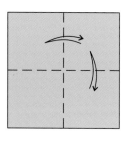

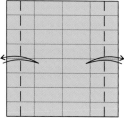

(5) Fold and unfold the edges in to the adjacent creases.

(6) Turn the model over left to right.

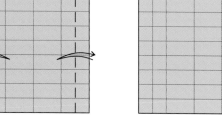

(7) Make a series of diagonal folds between the creases made previously.

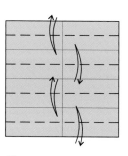

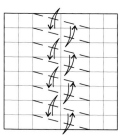

(8) Fold and unfold between the creases made previously, dividing each outer side into four sections.

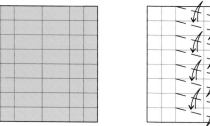

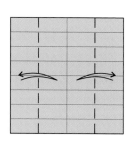

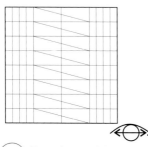

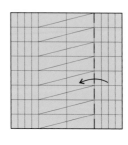

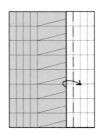

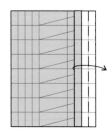

9 Turn the model over left to right.

10 Fold the right into the middle.

11 Fold the outer edge over along the crease made previously as shown.

12 Fold the section back over to the right.

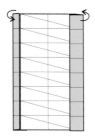

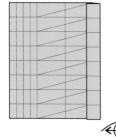

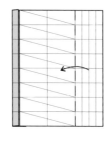

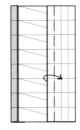

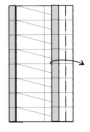

13 Turn the model over left to right.

14 Fold the right side into the middle.

15 Fold the edge over as shown.

16 Fold the section back again.

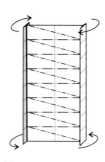

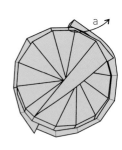

17 Fold the outer sections out to be perpendicular to the inner section.

18 Refold the creases made in the inner section. This will cause the model to twist (see opposite page).

19 Complete the twist to flatten the inner section.

20 The two sides should fit together. Now open out the folded edge (a) and wrap the paper around the edge of the opposite side.

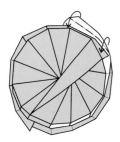 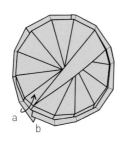 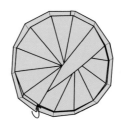 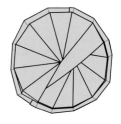

21) Fold the edge (a) from step 20 back over the inserted side of the model.

22) Open the paper (a) in the other end and wrap it around the edges at (b).

23) Fold the edge back over.

24) Complete.

FINAL ASSEMBLY OF THE DNA WHEEL

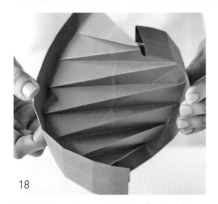

18
Refold the diagonal folds. This will cause the paper to start twisting.

19
Keep twisting the center section until the ends match up.

20
Unfold the paper in one end and insert the other end inside.

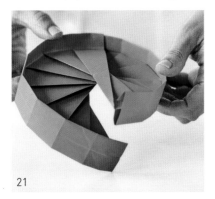

21
Flatten the center section and align the opposite ends.

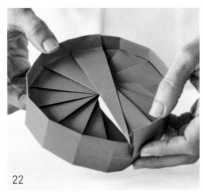

22
Open the paper in one of the ends and insert the other end.

23
Firm up the creases around the edges to complete the model.

Flexagon One

✳✳

The flexagon is a model that can be rotated in on itself. It is made from
six triangular tetrahedra that can be twisted to show four separate hexagonal faces.

7 × 7 in. (18 × 18 cm)   x 3

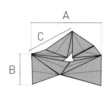

A   5 ½ in. (13.8 cm)
B   2 ⅜ in. (6 cm)
C   5 ½ in. (13.8 cm)

THE MODEL IS MADE FROM THREE SQUARES OF EQUAL SIZE. FOR EACH
OF THE THREE UNITS START WITH A SQUARE, COLORED SIDE UP.

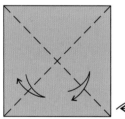

(1) Fold the square in half
diagonally. Then turn the
paper over left to right.

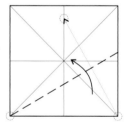

(2) Fold and unfold the paper
lengthwise along the middle.

(3) Fold the bottom right
corner up to touch the
center crease.

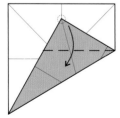

(4) Fold the corner back to the
folded edge.

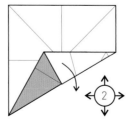

(5) Unfold the folded corner.

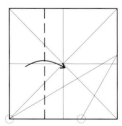

(6) Fold the left side over. The
edge should touch the
crease made previously.

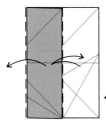 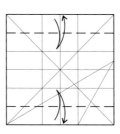 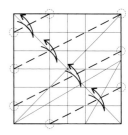

7 Fold the right side over. Then unfold both sides.

8 Fold in the top and bottom edges as shown to form thirds, then unfold.

9 Fold the edges in so they touch the adjacent creases, then unfold.

10 Make diagonal folds between the reference points indicated.

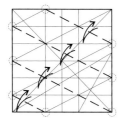 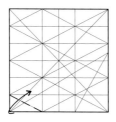 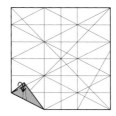 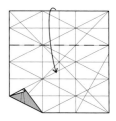

11 Make diagonal folds going in the opposite direction.

12 Fold over the bottom left corner.

13 Fold over the edge of the folded corner.

14 Fold top section down along the crease made previously.

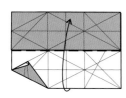  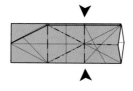 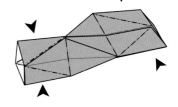

15 Fold the bottom third of the square up.

16 Fold and unfold the sides along the creases made previously.

17 Squash the model, reversing the creases and making a 3-D shape.

18 Squeeze the outer ends together to form more 3-D tetrahedron shapes.

  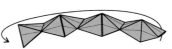 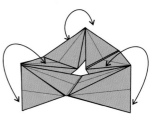

19 Make two more identical component parts (steps 1 to 18).

20 Insert one unit inside the other.

21 Insert the outer ends into each other to complete the ring.

22 The model is complete. Twist the sections to rotate the model.

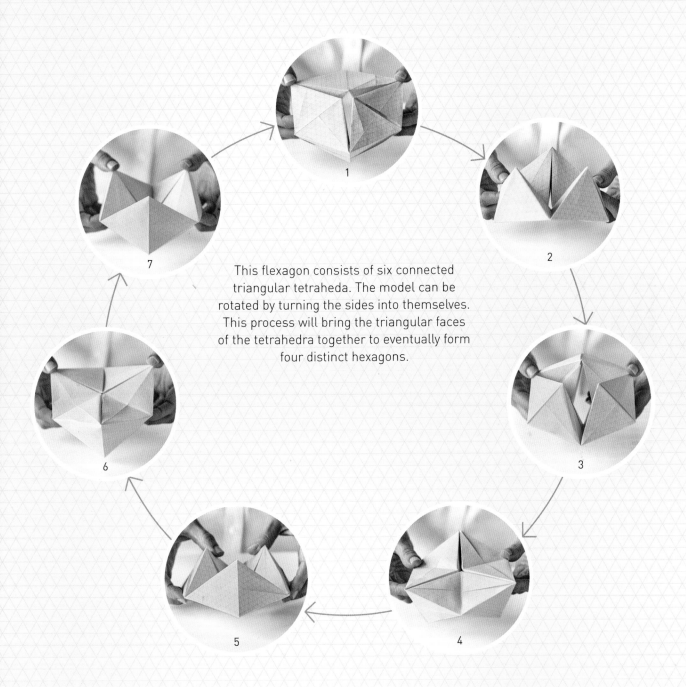

This flexagon consists of six connected triangular tetraheda. The model can be rotated by turning the sides into themselves. This process will bring the triangular faces of the tetrahedra together to eventually form four distinct hexagons.

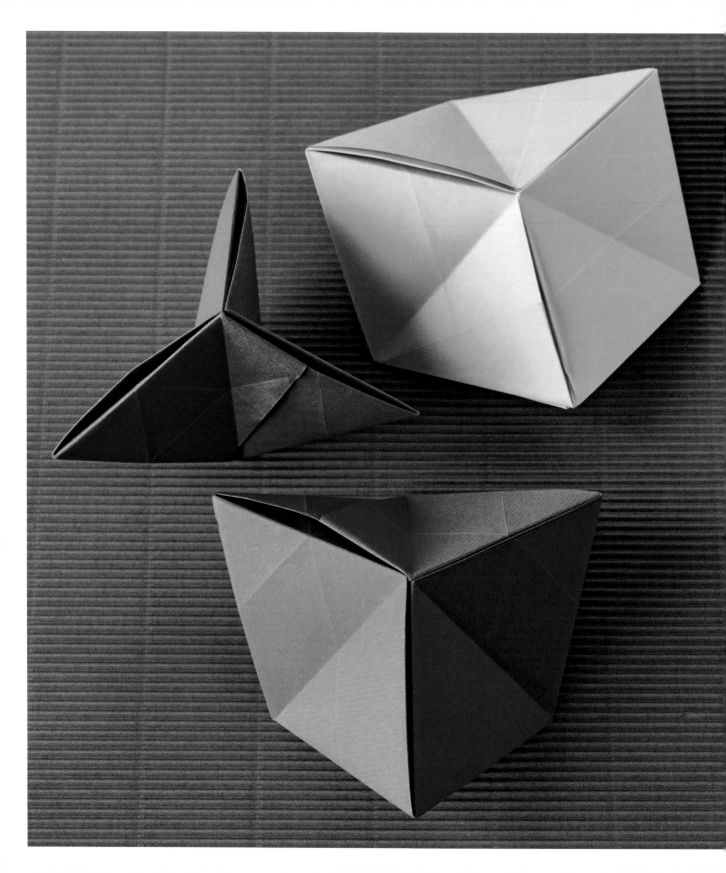

Flexagon Two
**

The second flexagon shares the same properties as the first flexagon in that it can be rotated. However, it is made from a series of connected equilateral triangles.

3½ × 7 in. (9 × 18 cm)   x 1

A   2 ¾ in. (6.9 cm)
B   2 ¾ in. (6.9 cm)
C   ¼ in. (0.5 cm)

START WITH A 2 × 1 RECTANGLE, COLORED SIDE DOWN.

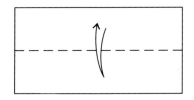

1 Fold and unfold the rectangle in half lengthwise.

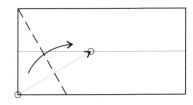

2 Fold the bottom left corner up to touch the middle crease. The fold should start from the opposite corner.

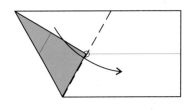

3 Fold the outer section over the edge of the folded corner.

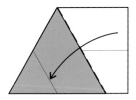

4 Fold the right side over along the edge.

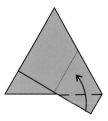

5 Fold the bottom corner up so it is level with the folded edge beneath.

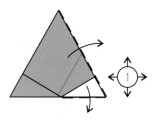

6 Unfold the model to step 1.

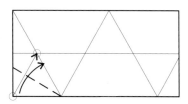

7 Fold the outer corner in to align the edge with the adjacent crease.

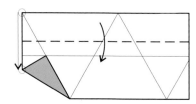

8 Fold the top edge down to align with the outer point of the folded corner.

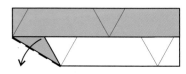

9 Unfold the corner folded in step 7.

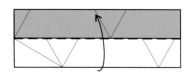

10 Fold the bottom edge up.

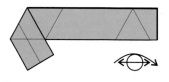

11 Fold the left side over along the crease made previously.

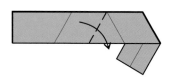

12 Fold the opposite side up to be parallel to the adjacent folded section.

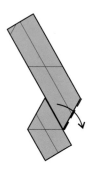

13 Undo the fold from the previous step.

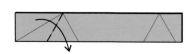

14 Turn the model over, left to right.

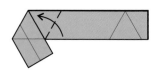

15 Fold the left section over so it is aligned with the adjacent crease.

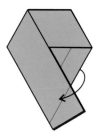

16 Move the bottom layer to the front above the folded section.

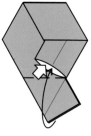

17 Fold the bottom section up and tuck one end of the model into the other.

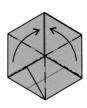

18 Fold the corners together to complete.

ROTATING THE FLEXAGON

(a) Hold the top corners and pull them apart.

(b) Continue opening out the hexagon until flat.

(c) Then push up the middle.

(d) Bring the inner point up and fold back to step (a).

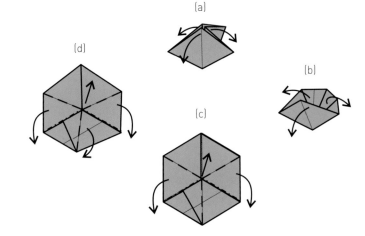

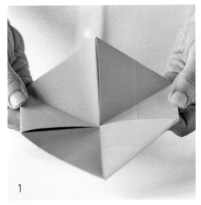

1

Fold the edges down, causing the middle of the hexagon to rise up.

2

Continue folding the edges downward and bring the triangular faces together.

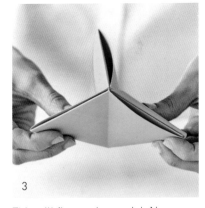

3

This will flatten the model. Now move your finger to the top point. (This was the middle of the hexagon.)

4

Holding the bottom edges together, pull apart the top tips of the equilateral triangles.

5

Ease the triangular faces apart while holding the bottom points together.

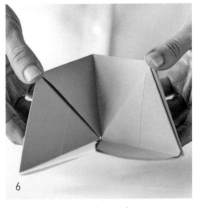

6

Open the face back to a hexagon (step 1).

Twist Star

**

The twist star is built around a folding process that twists together the two sides of a square. This rotation fold will form a star.

7 × 7 in. (18 × 18 cm)   x 1

A
B
C

A   5¼ in. (13.2 cm)
B   5¼ in. (13.2 cm)
C   ⅛ in. (0.12 cm)

START WITH A SQUARE, COLORED SIDE UP.

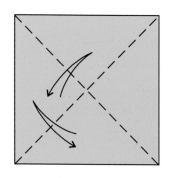

(1) Fold and unfold the square in half diagonally. Then turn the paper over left to right.

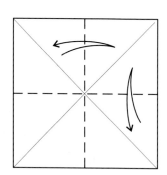

(2) Fold and unfold the square in half lengthwise along both axes.

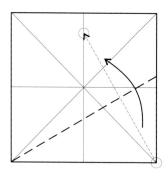

(3) Fold the bottom right corner up to touch the vertical middle crease.

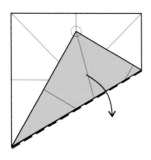

4 Undo the fold from the previous step.

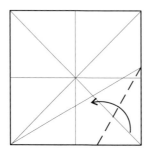

5 Fold the corner in to touch the crease made previously.

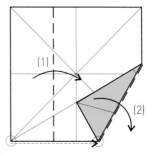

6 (1) Fold the left side in. Then (2) unfold the corner folded in the previous step.

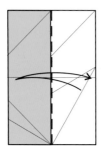

7 Fold the right side in and then out again.

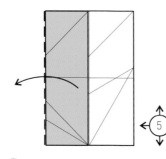

8 Unfold back into a square.

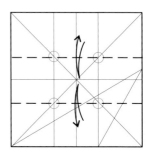

9 Fold the top and bottom edges in where the vertical creases touch the diagonal creases, to divide into thirds.

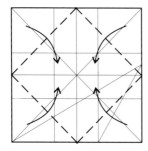

10 Fold the corners into the center of the paper.

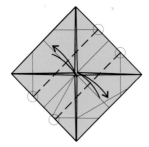

11 Fold the outer edges in at the points where the diagonal creases touch the edges, dividing the model into thirds, then unfold.

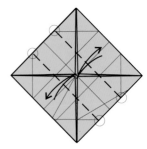

12 Repeat step 11 on the other axis.

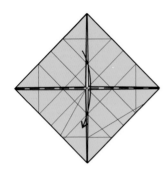

13 Fold the model in half.

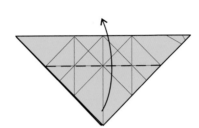

14 Fold the top layer up between the points where the diagonal creases touch the folded edge.

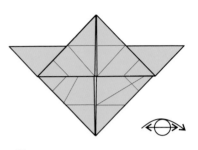

15 Turn the model over left to right.

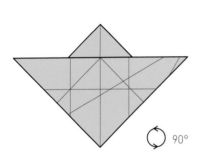

16 Rotate the model anticlockwise 90°.

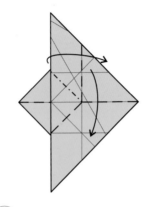

17 Fold the top section down. At the same time fold the top layer of the top section to the right. Hold the bottom left edge in place.

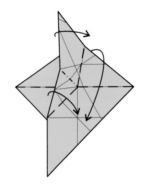

18 Fold in progress.

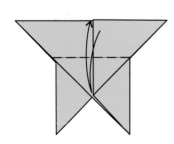

19 Fold the bottom section up and down again.

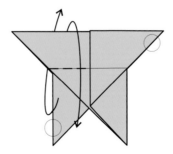

20 Fold the top left edge down while holding the right side still. This will cause a twist fold.

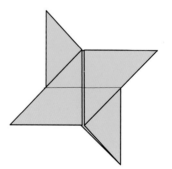

21 Complete.

Diamond Wreath

**

The diamond is made by applying the twist fold process to a diagonally folded square. The final diamond model has both points and pockets and can be used as a module to create larger constructions.

7 × 7 in. (18 × 18 cm)

× 16

A

B

C

A	⅛ in. (0.12 cm)
B	6 ⅝ in. (16.8 cm)
C	2 ⅞ in. (7.2 cm)

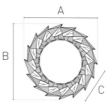
× 16

A

B

C

A	2 ⅞ in. (7.2 cm)
B	6 ⅝ in. (16.8 cm)
C	⅛ in. (0.12 cm)

START EACH OF THE 16 DIAMONDS WITH A SQUARE, COLORED SIDE UP.

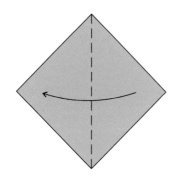

(1) Fold the square in half diagonally.

(2) Fold the model in half again.

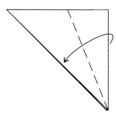

(3) Fold the right edge in diagonally to touch the opposite edge.

(4) Fold the top section up, separate the layers and squash it flat.

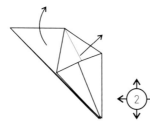
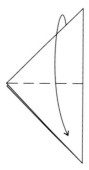

(5) Unfold the section folded previously (back to step 2).

(6) Fold the top layer of the bottom left side in diagonally to touch the opposite side, and unfold.

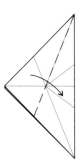

(7) Fold the top layer of the top left edge in diagonally to touch the opposite side.

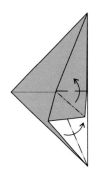

(8) Fold the middle section up and refold the bottom section along the creases made previously.

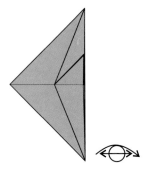

(9) Turn the model over left to right.

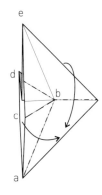

(10) Fold the top section down. At the same time fold the section (a–b–c) to the right. The section (b–d–e) should be held together with the top section.

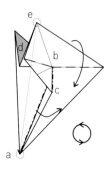

(11) This shows the fold in progress. When complete, rotate the model slightly.

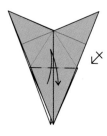

(12) Fold the bottom corner up and down. Repeat on the reverse.

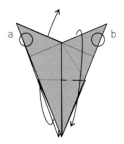

(13) Hold model at (a) and (b). Fold the right section (b) down while holding the left section (a) still. This twists and opens the model.

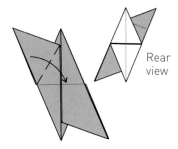

Rear view

(14) The twist is complete. Fold the left corner to the right.

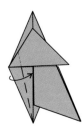

(15) Fold the left edge of the bottom section in to touch the center.

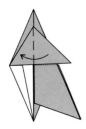
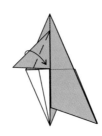
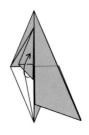

(16) Fold the top right corner back.

(17) Fold the corner into the adjacent edge.

(18) Fold the bottom corner up.

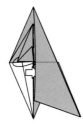
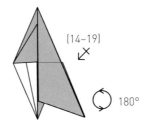
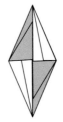

(19) Fold the edge of the top section over and into the adjacent pocket. When repeating this step, fold the edge into the bottom pocket.

(20) Rotate the model 180°. Repeat steps 14 to 19 on the other side.

(21) Diamond complete.

SIXTEEN UNITS CAN BE JOINED TOGETHER TO FORM A RING

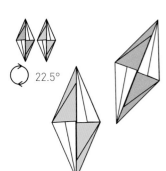
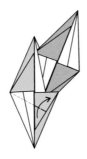
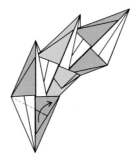
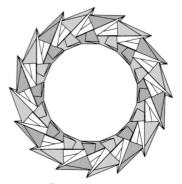

(1) Rotate one unit 22.5° and insert the bottom corner into the pocket in the top white section of the next unit.

(2) Fold up the point in the bottom section of the yellow diamond to link the diamonds together. Then add another unit.

(3) Fold over the bottom corner to lock the two units together. Then add more units repeating the linking process.

(4) Repeat this process to join 16 units together to form a wreath.

MODULAR PROJECTS

Crystal

*

The crystal explores a potentially infinite modular construction: each unit provides an opportunity to add two more units. As you develop the model think about complementary and contrasting color choices for the additional components.

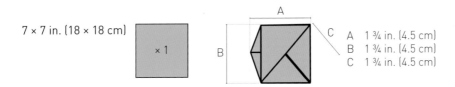

7 × 7 in. (18 × 18 cm)

× 1

A
B
C

A 1 ¾ in. (4.5 cm)
B 1 ¾ in. (4.5 cm)
C 1 ¾ in. (4.5 cm)

START WITH A SQUARE, COLORED SIDE UP.

1 Fold and unfold the square in half lengthwise.

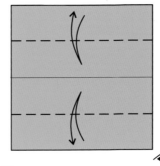

2 Fold the edges to touch the middle crease and unfold. Then turn the model over left to right.

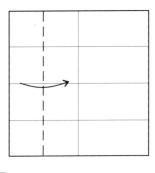

3 Fold the left edge in to touch the middle crease.

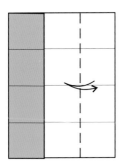

4 Fold and unfold the edge in to touch the middle crease.

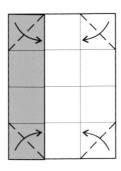

5 Fold all the corners in.

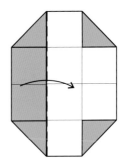

6 Fold one side in.

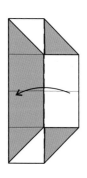

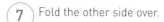

(7) Fold the other side over.

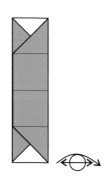

(8) Turn the model over left to right.

(9) Fold and unfold along the middle.

(10) Fold the top and bottom edges into the middle.

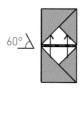

60°

(11) Tuck one end into the other. This process shows the two sides sliding into each other.

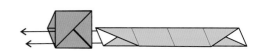

(12) Make another module. Unfold it to step 8 and slide the flat version, open side up, into a completed unit.

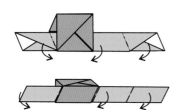

(13) Fold the edges of the corners down and slide one end into the other.

(14) Push the two ends together. Add more units to complete the crystal.

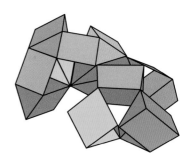

(15) Complete.

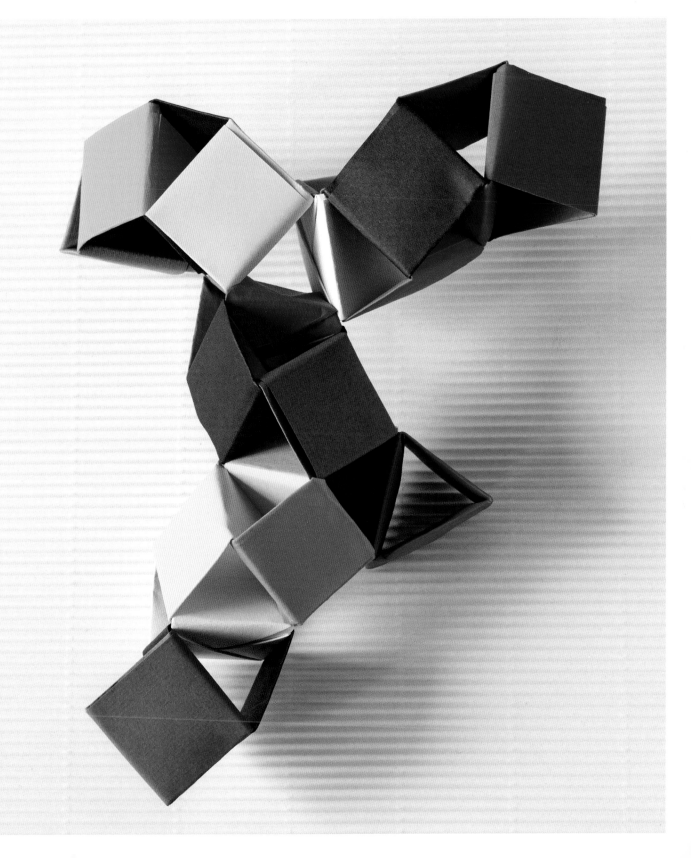

Octahedron Nolid

This model is a two-piece assembly. Each unit is made in the same way,
with one being rotated to fit into the base of the other.

7 × 7 in. (18 × 18 cm)

×2

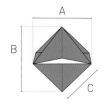

A   3 ½ in. (9 cm)
B   5 in. (12.6 cm)
C   3 ½ in. (9 cm)

THE MODEL IS MADE FROM TWO SQUARES. FOR EACH UNIT START WITH A SQUARE, COLORED SIDE UP.

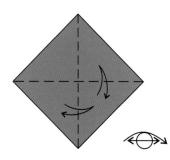

(1) Fold and unfold the square in half diagonally along both axes. Then turn the paper over left to right.

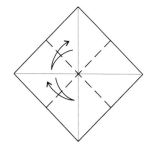

(2) Fold and unfold the square in half lengthwise.

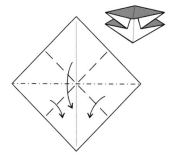

(3) Fold the top half down. At the same time, fold in the sides to make a preliminary base.

(4) Rotate the model 180°.

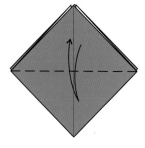

(5) Fold and unfold the top layer along the middle.

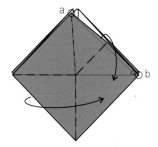

(6) Fold the left side to the right to be perpendicular to the rear section. At the same time, fold down the top section, causing (a) to touch (b).

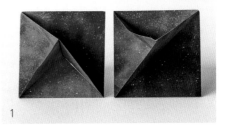

1

Two similar units.

2

Align the points on one unit with the base.

3

Slide the two units together.

4

Model complete.

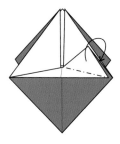

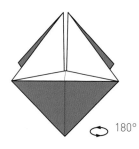

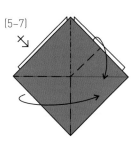

(5–7)

(7) Fold the corner behind and into the pocket behind.

(8) Rotate the unit 180° to work on the reverse.

180°

(9) Repeat steps 5 to 7.

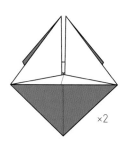

×2

(10) One unit is complete. Make a second unit in the same way to complete the model.

(11) Turn one unit upside down and rotate it 90°.

(12) Line up the two units and insert one into the other (see previous page, right).

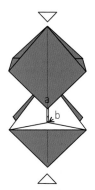

(13) First line up the points and tuck the corner (a) beneath the edge (b), front and back. Then slide one unit into the other.

(14) Continue pushing the units together.

(15) Complete.

Dodecahedron

✱✱

The dodecahedron is one of the platonic solids, constructed from twelve regular pentagons.
In ancient Greece its shape was seen as fundamental to the arrangement of the universe.

7 × 7 in. (18 × 18 cm)   × 12

A   5 ⅔ in. (14.4 cm)
B   5 ⅔ in. (14.4 cm)
C   5 ⅔ in. (14.4 cm)

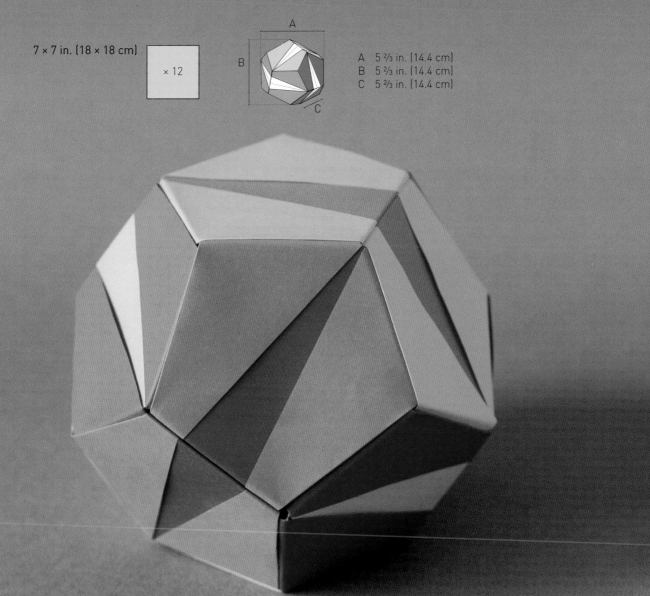

THE MODEL IS MADE FROM 12 UNITS. FOR EACH UNIT START WITH A SQUARE, COLORED SIDE UP.

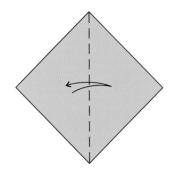

(1) Fold and unfold the square in half lengthwise.

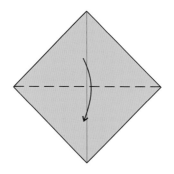

(2) Fold the square in half.

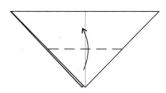

(3) Fold the top layer of the bottom corner up to the middle of the folded edge above.

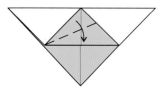

(4) Fold the corner down to the adjacent folded edge.

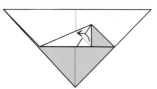

(5) Fold the corner up and align the outer edge with the folded edge.

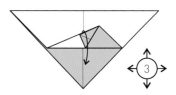

(6) Unfold to step 3.

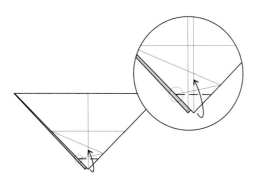

(7) Fold the the bottom corner up. Note the reference point as the end of the crease made previously.

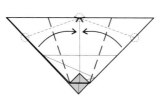

(8) Fold the outer edges in. The creases start from the tip of the folded triangle and will cause the edges to touch the middle of the top folded edge.

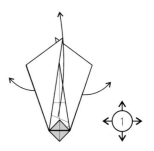

(9) Unfold back into a square, but with the bottom corner still folded in.

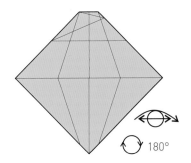

10 Turn the model over left to right and rotate 180°.

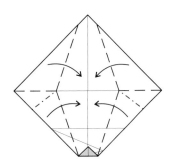

11 Fold in the outer edges along the creases made previously.

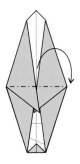

12 Fold the top section to the back.

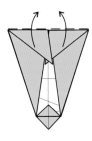

13 Fold the corners up.

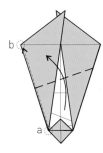

14 Fold the bottom section up, causing the bottom corner (a) to touch the top corner (b).

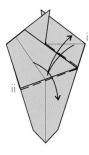

15 (i) Fold and unfold the top right corner over the adjacent edge. (ii) Then fold the section back.

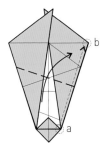

16 Fold the other side up so the bottom corner (a) touches the top corner (b).

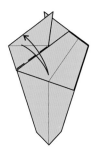

17 Fold and unfold the top left corner over the adjacent edge.

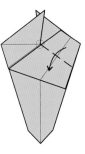

18 Fold the end of the top section in. The fold should start from the point where the paper touches the folded edge beneath.

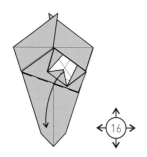

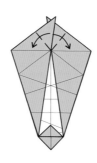

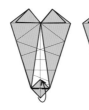

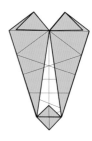

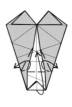

(19) Unfold the folded section to step 16.

(20) Fold the top corners out to align the inside edges with the horizontal creases.

(21) Module complete.

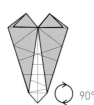

(22) Make a second module in the same way.

(23) On one module, fold the tip up; fold it down on the other.

(24) Open up the end of the left module.

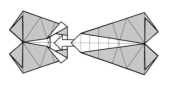

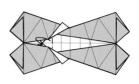

(25) Rotate the two modules by 90° to point the tip of one into the opening in the other.

(26) Insert one module inside the other.

(27) Fold the tip of the module being inserted behind to lock the two modules together.

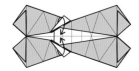

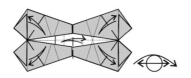

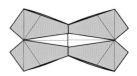

(28) Fold the paper on the receiving module back to complete the link.

(29) Crease the model around the edges of the pentagonal shapes; these form the faces of the dodecahedron. Turn model over left to right.

(30) Make 10 more modules and link them together in pairs.

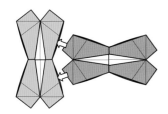

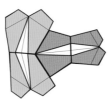

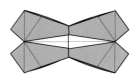

(31) Insert the tabs on one pair of linked modules into the pockets of another pair.

(32) Two pairs of linked modules connected.

(33) The final model is made from six modules in complementary colors.

LINKING THE MODULES TO FORM A DODECAHEDRON

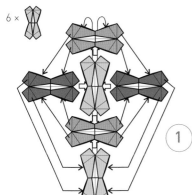

6 ×

(1) The final dodecahedron is made from a series of six linked modules.

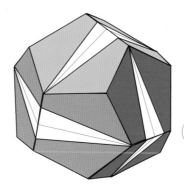

(2) Complete.

Truncated Cube
**

This is a cube with the corners removed. The model is made from six similar units that fit neatly together. The final model is formed by a tesselation of squares and triangles.

7 × 7 in. (18 × 18 cm)   × 6

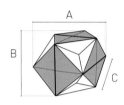

A   3 ⅓ in. (8.4 cm)
B   3 ⅓ in. (8.4 cm)
C   3 ⅓ in. (8.4 cm)

THIS MODEL IS MADE FROM SIX EQUAL-SIZED UNITS.
FOR EACH UNIT START WITH A SQUARE, COLORED SIDE UP.

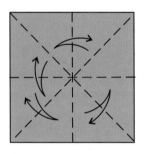

1 Fold and unfold the square in half lengthwise and diagonally along all axes.

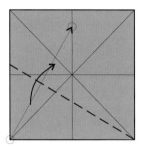

2 Fold the bottom left corner up to touch the vertical middle crease.

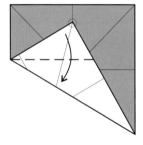

3 Fold the corner down so that its edge touches the adjacent folded edge.

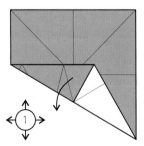

4 Unfold the corner.

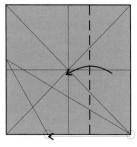

5 Fold the right edge into the crease made previously.

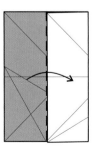

6 Fold the edge over again.

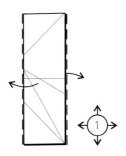

7 Unfold back into a square.

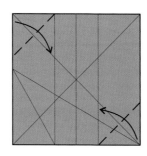

8 Fold the top left and bottom right corners in to touch the creases made previously.

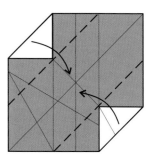

9 Fold the edges in again.

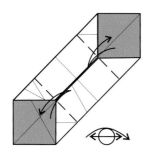

10 Fold and unfold where indicated. Then turn the model over left to right.

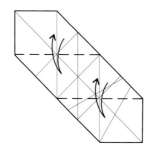

11 Fold and unfold the paper where indicated.

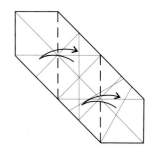

12 Fold and unfold the paper where indicated.

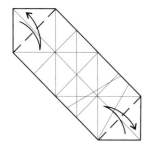

13 Fold the corners in and out again.

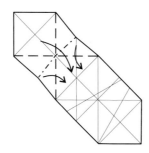

14 Fold the outer section in along the creases made previously.

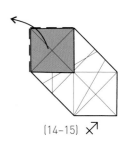

(14–15)

15 Unfold the section. Then repeat steps 14 to 15 on the other end to complete one of the units. Make five more.

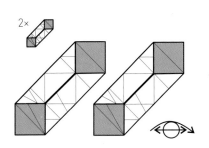
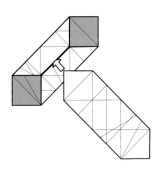
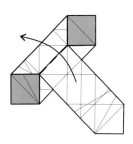

16 Start with two units, then turn one of the units over left to right.

17 Insert the corner of one unit into a pocket in the other.

18 Fold the unit back over.

ADDING A THIRD UNIT

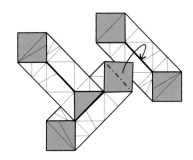
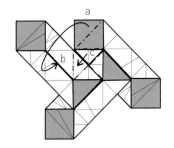
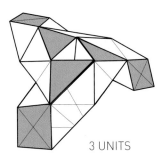

19 The principle is that each corner will fold into a pocket. This shows a corner folding into a third unit.

20 Tuck the corner (a) into the pocket (b). This will cause the layer (c) to fold over the adjacent section in unit (b).

21 This shows three units combined.

3 UNITS

SIX-UNIT ASSEMBLY

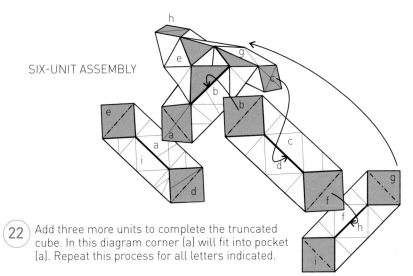

22 Add three more units to complete the truncated cube. In this diagram corner (a) will fit into pocket (a). Repeat this process for all letters indicated.

23 Complete.

Hexahedron

✳✳

This modular project makes a series of regular solids that can be made from similarly folded units. The units fit together by inserting the tab of one unit into a pocket in another. The hexahedron is made up of three units.

7 × 7 in. (18 × 18 cm)

×3

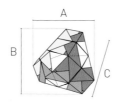

A   3 ½ in. (9 cm)
B   3 ½ in. (9 cm)
C   2 ⅜ in. (6 cm)

THE MODEL IS MADE FROM THREE UNITS. FOR EACH UNIT START WITH A SQUARE, COLORED SIDE UP.

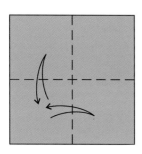

(1) Fold and unfold the square in half lengthwise along both axes.

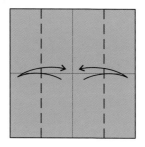

(2) Fold and unfold the edges into the middle.

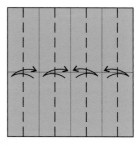

(3) Fold and unfold between the creases made previously.

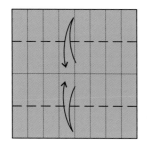

(4) Fold and unfold the top and bottom edges into the middle.

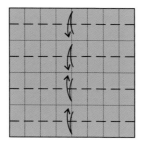

(5) Fold and unfold between the creases made previously. Then turn the model over left to right.

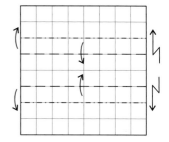

(6) Fold the outer edges in and out again making two zigzag folds.

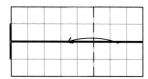

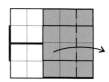

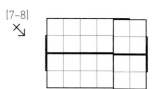

7 Fold one side over.

8 Fold the section back again where indicated.

9 Repeat steps 7 to 8 on the other side.

10 Fold the outer corners in.

11 Fold the edges over again.

12 Turn the unit over left to right.

a

b

c

HEXAHEDRON ASSEMBLY

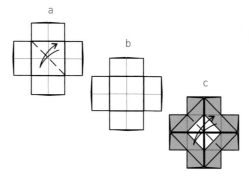

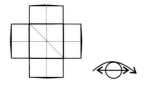

13 Fold the outer edges in and out again to make creases that will be used in the assembly process.

14 Each model will require a different crease.
(a) Diagonally on the reverse for the hexahedron.
(b) No crease for the cube.
(c) Diagonally on the front for more advanced assemblies.

15 Turn the unit over left to right.

JOINING TWO UNITS

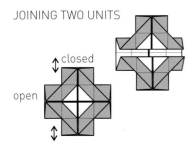

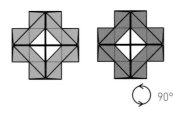

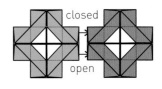

(16) The unit has an open and a closed axis. The objective of the assembly is to place the open tabs into the closed tabs.

○ 90°

(17) Align two units, then rotate one 90°. This will match the "closed" axis with the "open" axis.

(18) Insert the closed tab into the open tab.

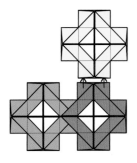

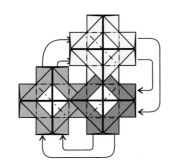

(19) Add a third unit, again inserting a closed tab into an open tab.

(20) Three units can be joined to make a hexahedron.

(21) Complete.

ASSEMBLING LARGER MODELS

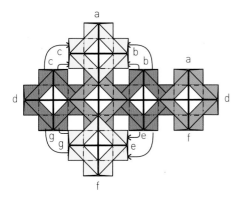

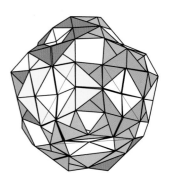

(22) Six units can be assembled into a cube. The net above shows the assembly. Insert tab (a) into pocket (a). Then repeat for (b) to (g).

(23) Cube complete.

(24) The module can be used to make other regular solids by combining 12 units.

ADVANCED
MODULARS

Equilateral Module
✳✳

This modular project enables the construction of polygons made from equilateral triangles. There are various methods of weaving units together. Unlike most unit-based origami projects, some of these assemblies include both similar and mirrored units.

7 × 7 in. (18 × 18 cm) ICOSAHEDRON

×5

×5

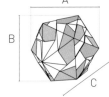
× 10

10 units

A

B

C

A   6 ⅜ in. (16.2 cm)
B   6 ⅜ in. (16.2 cm)
C   6 ⅜ in. (16.2 cm)

OTHER ASSEMBLIES

×2
×2

4 units

×5

5 units

THE ICOSAHEDRON IS MADE FROM 10 UNITS. FIVE ARE THE MIRROR IMAGE OF THE OTHER FIVE. START WITH A SQUARE, COLORED SIDE UP.

1 Fold and unfold the square in half lengthwise. Then turn the module over left to right.

2 Fold the bottom edge up to touch the middle crease.

3 Fold the top left corner down to touch the folded edge. Then unfold back into a square.

4 Fold the top right corner into the center of the bottom section.

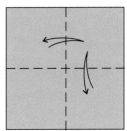

5 Fold the bottom left corner up to touch the crease in the folded edge above.

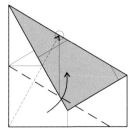

6 Fold the corner over at the points where the top and bottom folded sections touch.

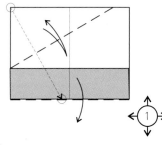
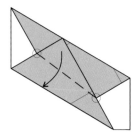

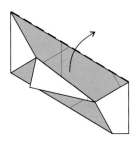

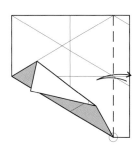

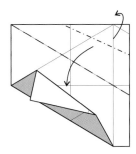

(7) Fold the top section out.

(8) Fold the right edge in at the end of the folded section.

(9) Fold the top right corner down along the crease made previously, then back up.

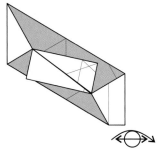

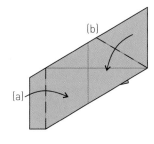

(b)

(a)

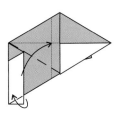

(10) Turn the model over left to right.

(11) Fold the outer edge (a) in along the crease made previously. Then fold in the top right corner across (b).

(12) Fold the bottom corner in. Then fold the bottom section up to align the section edge with the edge of the other folded triangle above.

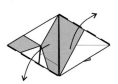

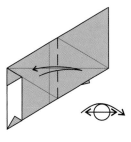

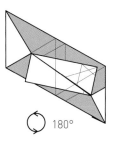

180°

(13) Unfold both sides.

(14) Fold and unfold along the middle. Then turn the model over left to right

(15) Rotate the model 180°.

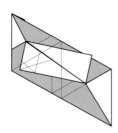

(16) Module complete. The 10-unit assembly requires five similar modules and five modules that mirror the module above.

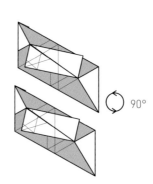

(17) Make five similar modules. Start with two and rotate one 90°.

90°

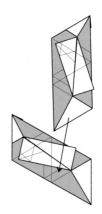

(18) Place the corner of one module over the triangular section of the second module.

(19) Fold the corner of the vertical module behind and beneath the folded edge of the horizontal module.

(20) Fold the edge of the corner behind to lock the two units together.

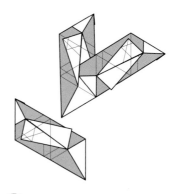

(21) Rotate the joined modules and add a third and fourth module. Repeat the connecting process.

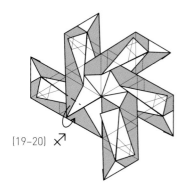

(19–20)

(22) Add a fifth unit and repeat steps 19 to 20 to link it.

(23) Insert the triangular point of the fifth unit into the matching space on the first unit, then repeat the linking process.

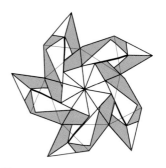

(24) Five-unit assembly. This can be made into a 10-sided shape by turning the model over and joining the triangular points to a mirror image.

TO COMPLETE THE ICOSAHEDRON MAKE FIVE MORE MODULES THAT SHOULD
MIRROR THE FIRST FIVE. FOR EACH OF THE FIVE UNITS START WITH A SQUARE,
COLORED SIDE UP, AND REPEAT STEPS 1 TO 2 OF THE STANDARD MODULE.

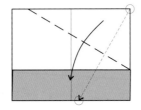

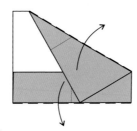

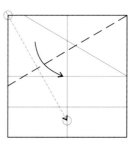

(1) Fold the top right corner down
to touch the folded edge. This
is the opposite corner to the
first unit.

(2) Unfold back into a square.

(3) Fold the top left corner into
the center of the bottom
section.

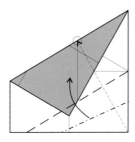

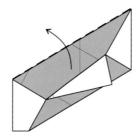

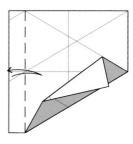

(4) Fold the bottom corner up
to touch the crease in the
folded edge above. Then fold
the section back again.

(5) Fold the top section out.

(6) Fold the left edge over at
the point where the bottom
fold starts, and then unfold
it again.

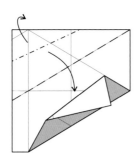

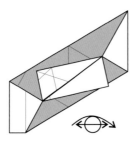

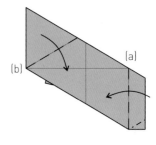

(7) Fold the top left corner
down along the crease
made previously.

(8) Turn the model over left
to right.

(9) Fold the outer edge in
along the crease (a) made
previously. Then fold in the
opposite side and fold in the
bottom corner (b).

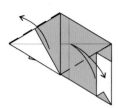

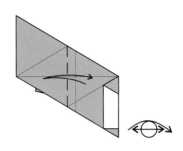

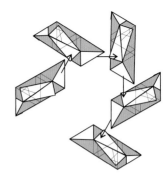

(10) Fold and unfold the bottom corner, then fold the top section out.

(11) Fold and unfold along the center. Then turn over left to right.

(12) Assemble the five modules in a similar way to the first five (see page 107).

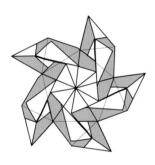

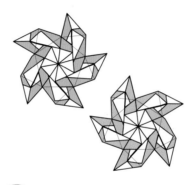

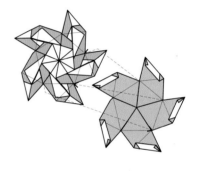

(13) Module complete.

(14) To form the icosahedron, start with two modules, one a mirror image of the other. Turn one over.

(15) Align the two modules and bring them together.

(16) Now weave the two assemblies together. Flatten the equilateral triangles from one assembly onto the matching triangles on the other.

(17) Fold the corners of the triangles behind and into the pockets formed by the folded edges beneath.

(18) Icosahedron assembly complete. Alternative assemblies can be made from four and five units.

Harlequin Module

✱✱

This is a module that can be used to make a series of shapes. The hexahedron is made from three units, although more regular shapes can be made from 1, 2, 3, 6, 12, and 30 units.

7 × 7 in. (18 × 18 cm)

×3

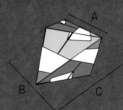

A   4 ¾ in. (12 cm)
B   3 ⅞ in. (9.75 cm)
C   4 ¾ in. (12 cm)

THE HEXAHEDRON IS MADE FROM THREE UNITS. FOR EACH UNIT START WITH A SQUARE, COLORED SIDE DOWN.

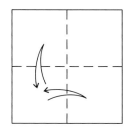

(1) Fold and unfold the square in half lengthwise along both axes.

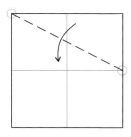

(2) Fold the top right corner down between the top left corner and right edge of the horizontal crease.

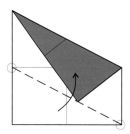

(3) Fold the bottom left corner up between the corner and the crease made previously.

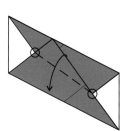

(4) Fold the center section down where the folded layers touch.

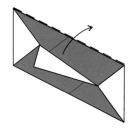

(5) Fold the inside corner of the top section back out again.

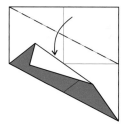

(6) Fold the corner back down again, along the crease made previously.

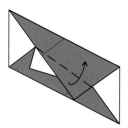

(7) Fold the corner up along the folded edge beneath.

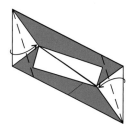

(8) Fold the outer edges in on both sides.

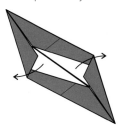

(9) Lift up the inner white section and bring them on top of the folded outer corners.

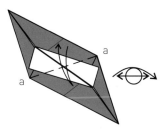

(10) Fold the model in half along the axes (a–a). Then turn the model over left to right.

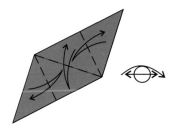

(11) Fold the corners in and out again. Then turn the model over left to right.

(12) Module complete.

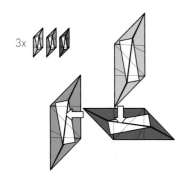

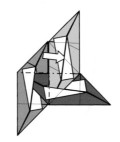

13 Three units can be assembled to make a hexahedron. Weave the modules together by inserting the corners into the pockets as indicated.

14 The final corner inserts into the pocket and the middle section becomes a triangular pyramid made from three connected isosceles triangles.

15 Turn the model over to look at the underside.

16 Fold one side in.

17 Fold the next section (yellow) up and tuck the corner into the adjacent pocket. Repeat this on the third section.

18 Hexahedron assembly complete.

MORE ASSEMBLIES

Three units

Four units

Five units

19 Six units can be assembled to make a cube.

20 Twelve units will make a stellated octahedron.

21 Thirty units can be assembled to make a stellated icosahedron.

Flower Ball

**

The flower ball unit can be assembled in a similar way to the previous project. The unit is made more intriguing by added folded layers that enhance its texture and overall look.

7 × 7 in. (18 × 18 cm)   × 30

A
B
C

A 7 in. (18 cm)
B 7 in. (18 cm)
C 7 in. (18 cm)

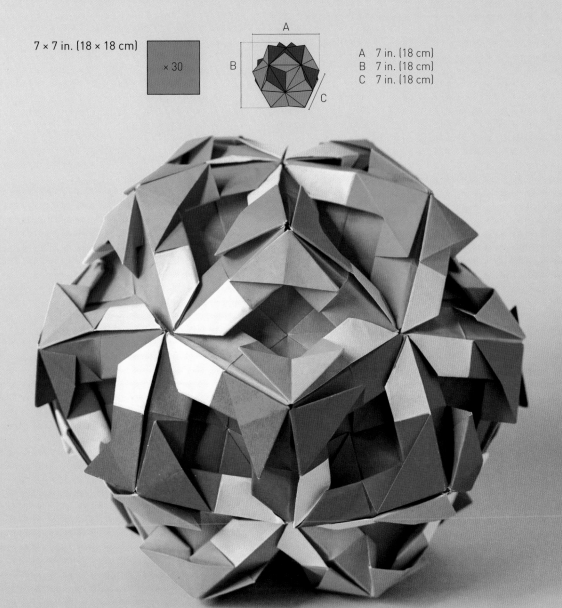

THE MODEL IS MADE FROM 30 UNITS. FOR EACH UNIT START WITH A SQUARE, COLORED SIDE DOWN.

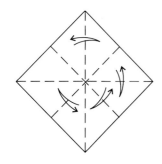

(1) Fold and unfold the square in half lengthwise and diagonally along all axes.

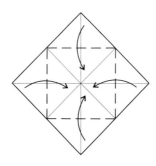

(2) Fold the corners into the center.

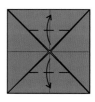

(3) Fold the top and bottom inside corners out to the folded edges.

(4) Fold and unfold the top left and bottom right corners into the center.

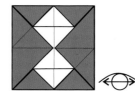

(5) Turn the model over left to right.

(6) Fold the top left and bottom right corners into the center.

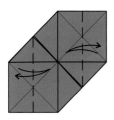

(7) Fold and unfold the edges into the center crease.

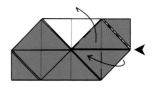

(8) Fold the top and bottom edges into the center crease.

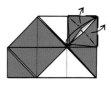

(9) Fold one side in and fold up the bottom edge to open up the section.

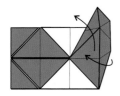

(10) Fold in progress.

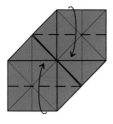

(11) Open the layers of the point and squash flat.

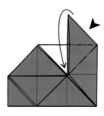

(12) Fold the corner out and open up the point.

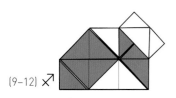

(9–12) ↗

13 Repeat steps 9 to 12 on the other side.

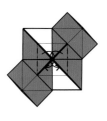

14 Fold and unfold the inner triangles out to the adjacent folded edge.

15 Fold the two corners behind and into the model.

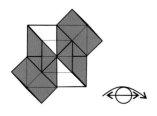

16 Turn the model over left to right.

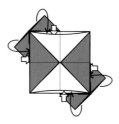

17 Fold the corners into the adjacent pockets.

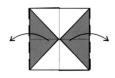

18 Fold the corners out.

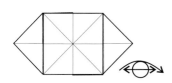

19 Turn the model over left to right.

+1

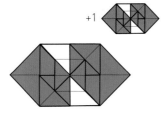

20 Unit complete. Add a second unit and rotate one 90°.

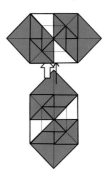

21 Insert the point of one unit into the pocket of the other.

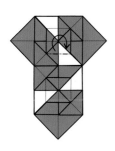

22 Fold the corner behind to connect the two units.

+1

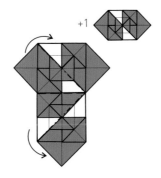

23 Fold the units diagonally and then add a third unit.

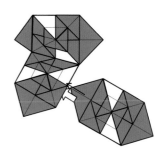

24 Insert the point into the pocket of the connected units.

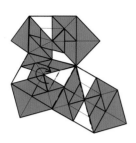

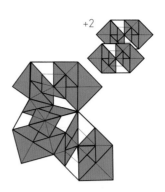

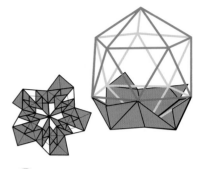

(25) Fold the corner behind to lock the modules together.

(26) Add two more units to complete the ring of five units.

(27) The five linked units form the base of the model. Turn it upside down and start to build the shape around an imaginary icosahedron.

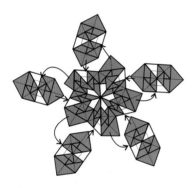

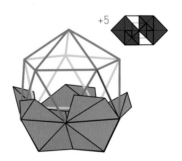

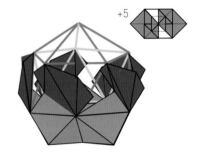

(28) Add five more units to build up the shape.

(29) This shows two layers of assembled units. Add a third layer of five units.

(30) Add another layer of five units.

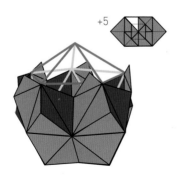

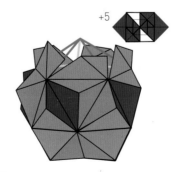

(31) Add another layer of five units.

(32) Add another layer of five units to complete the solid.

(33) Complete. Other assemblies can be made from 1, 2, 3, 6, and 12 units.

The flower ball is made from 30 units assembled as a stellated icosahedron (a 20-sided icosahedron with each face being replaced by a three-sided triangular pyramid). The units can be combined in other ways, with 3 units forming a tetrahedron, 6 units forming a cube, and 12 units making a stellated octahedron.

Vertex Module
✳✳

This modular project creates a unit that can be used to form the edge of a polygon. The module is quite flexible and can be used to make a variety of shapes. The 90-unit construction is made from a 3–D tesselation of regular hexagons and pentagons.

3½ × 1¾ in. (9 × 4.5 cm)   × 90

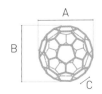

A   7 in. (18 cm)
B   7 in. (18 cm)
C   7 in. (18 cm)

THE TRUNCATED ICOSAHEDRONS, OR FOOTBALL-SHAPED ASSEMBLY, IS MADE FROM 90 UNITS. FOR EACH UNIT START WITH A 2 × 1 RECTANGLE, COLORED SIDE DOWN.

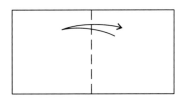

1 Fold and unfold the rectangle along the center.

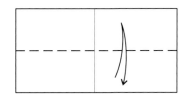

2 Fold and unfold along the center of the other axis.

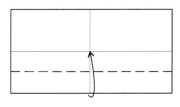

3 Fold the bottom edge up to the center crease.

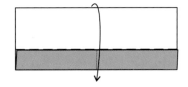

4 Fold the top edge over.

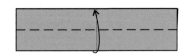

5 Fold the bottom edge up.

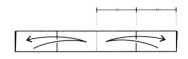

6 Fold and unfold the outer edges into the center.

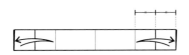
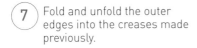

7 Fold and unfold the outer edges into the creases made previously.

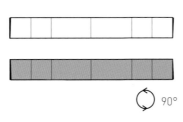

8 Make another unit, colored side up. Then rotate one 90°.

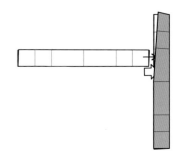

9 Insert one end between the layers of another unit.

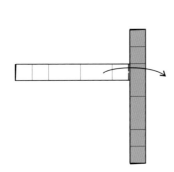

10 Fold the inserted unit over the other.

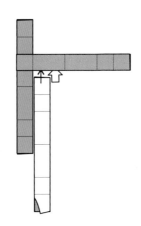

11 Add a third unit and insert it into the second unit.

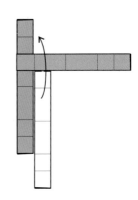

12 Fold the third unit up.

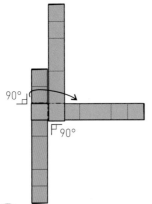

13 Fold the units perpendicular to one another, then fold the end of the first unit over the edge of the third.

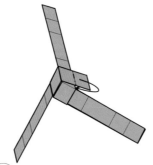

14 Fold the end of the first unit over and tuck the end between the layers of the third unit.

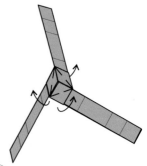

15 Make three diagonal creases and fold up each of the units up.

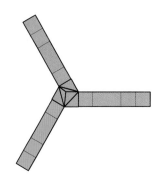

16 Three units are now assembled.

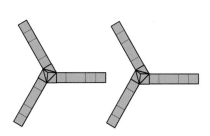

17 Repeat the assembly on three more units.

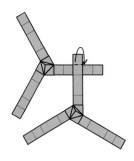

18 Place one assembly over the other and tuck the end of the tab behind and between the layers of the assembly underneath.

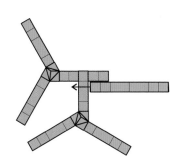

19 Add a new unit.

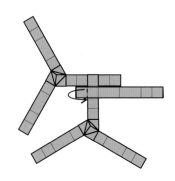

20 Fold the end of the unit behind and between the layers of the assembly underneath.

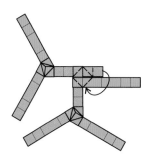

21 Fold the tip of the assembly over and tuck it into the adjacent unit. This repeats the assembly process steps 13 to 15.

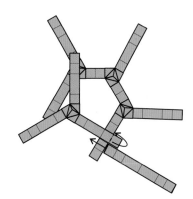

22 Add another assembly and an additional unit to connect all of the units together.

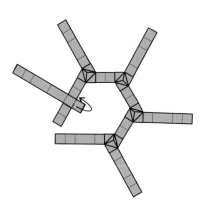

23 Add another unit to complete the 10-unit assembly.

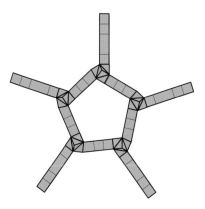

24 10-unit assembly complete.

DODECAHEDRON

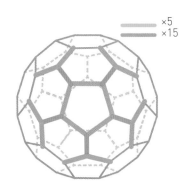

×10
×10
×10

1 The dodecahedron is made from 30 units. Start with a 10-unit assembly and make a pentagon from a single color.

2 Connect two more units to each of the outlying single units.

3 Complete the dodecahedron by adding additional units to the reverse of the model.

TRUNCATED ICOSAHEDRON

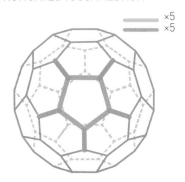

×5
×5

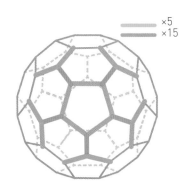

×5
×15

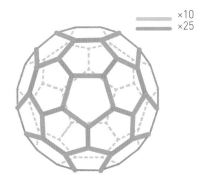

×10
×25

1 The truncated icosahedron can be made from 90 units. Start with a three-unit assembly.

2 Join two additional units to each of the outlying units.

3 Add an additional unit to complete hexagons. At each node add two additional units using the three-unit assembly process.

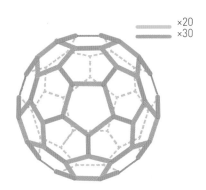

×20
×30

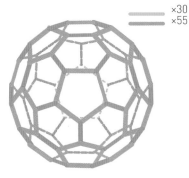

×30
×55

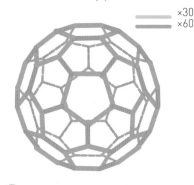

×30
×60

4 Continue adding units, the first to complete pentagons, with two additional units using the three-unit assembly process.

5 Continue adding units to the reverse, following the pentagon/hexagon assembly process.

6 The truncated icosahedron is complete.

Starburst

**

The starburst is assembled in a similar way to the previous projects, but the design of its units leads to extended stellations. These instructions are for a 30-unit starburst.

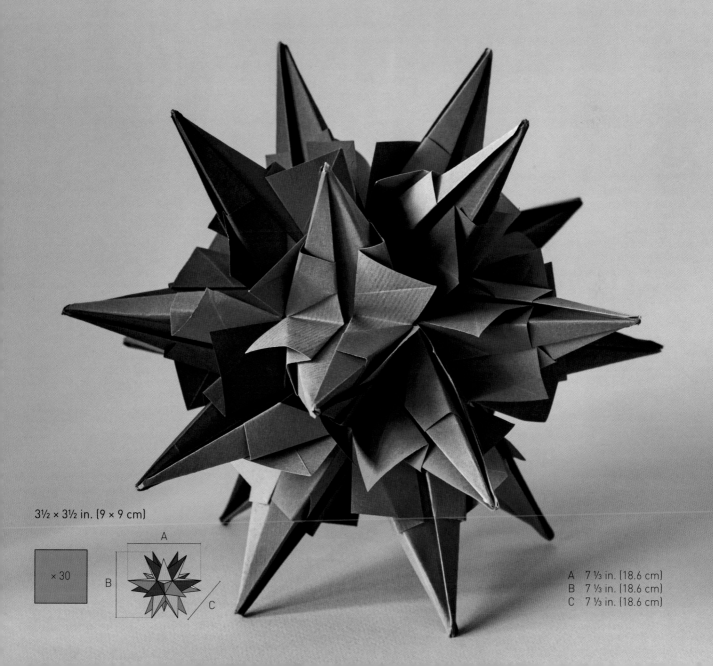

3½ × 3½ in. (9 × 9 cm)

× 30

A
B
C

A 7 ⅓ in. (18.6 cm)
B 7 ⅓ in. (18.6 cm)
C 7 ⅓ in. (18.6 cm)

THE STARBURST ASSEMBLY IS MADE FROM THIRTY UNITS. EACH UNIT STARTS WITH A SQUARE, COLORED SIDE UP.

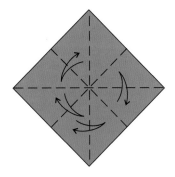

① Fold and unfold the square in half lengthwise and diagonally along all axes.

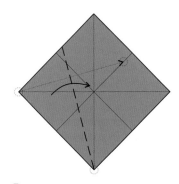

② Fold the left corner in to touch the diagonal middle crease. The fold starts from the bottom corner.

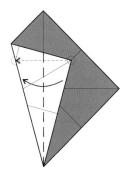

③ Fold the edge back out to the adjacent folded edge.

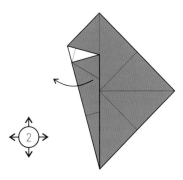

④ Unfold the model back into a square (step 2).

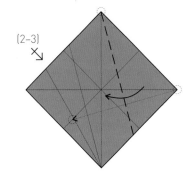

⑤ Fold the right corner into the opposite diagonal crease and repeat steps 2 to 3.

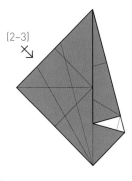

⑥ Repeat steps 2 to 3 on the other side.

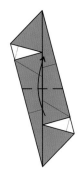

⑦ Fold the bottom section up.

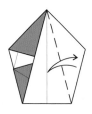

⑧ Fold and unfold the right side in.

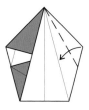

⑨ Fold the edge into the crease made in step 8.

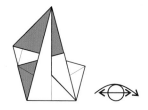

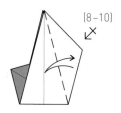

(8–10)

10 Fold the edge over again.

11 Turn the model over left to right.

12 Repeat steps 8 to 10.

Front view   Back view

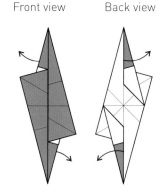

Front view   Back view

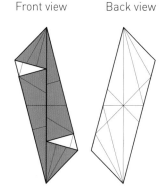

13 Fold the top section down.

14 Unfold the corners on both sides (back to step 7).

15 Unit complete.

LINKING TWO MODULES

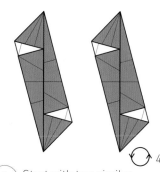

45°

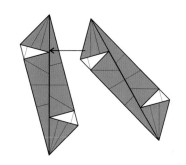

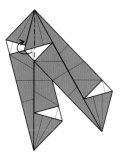

1 Start with two similar modules. Rotate one 45°.

2 Move the tab from one module over the edge of the other.

3 Fold the corner of the top unit behind and into a pocket in the unit beneath to link the two units.

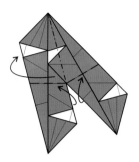

This shows the two linked units as part of a five-unit assembly.

4 Fold the assembly in half at the edge where the two units meet.

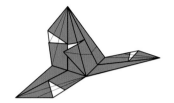

5 The two joined modules will look like this.

Three units as part of a five-unit assembly.

6 Add a third module. Repeat the linking process by tucking the corner of one unit into the pocket of the other.

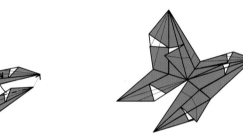

7 Three modules combined.

8 Working to join five modules together, you now need to add two more modules. Connect them in a similar way to the previously assembled units.

9 Five joined modules.

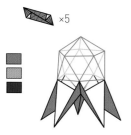

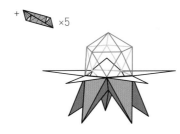

10 Start with five assembled units of a similar color.

11 Add five more modules of another color in between the modules that have been assembled previously.

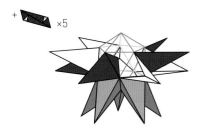

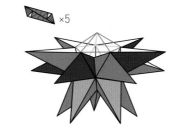

12 Add five more modules of a third color. These modules are all added to the right of the top edge of the assembly.

13 Add five more units.

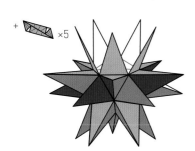

14 The next five units join up the units added in the previous two steps.

15 Complete. Other assemblies can be made from three, six, and twelve units.

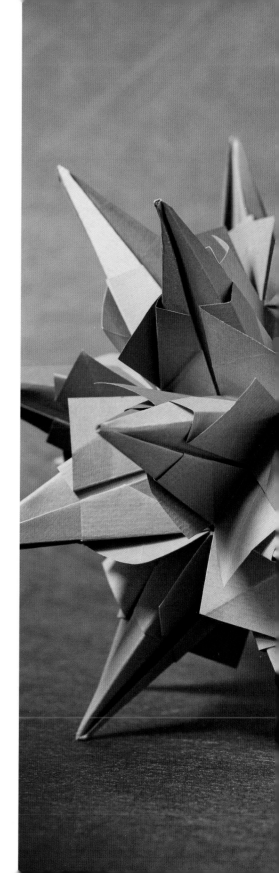

RESOURCES AND ACKNOWLEDGMENTS

AUTHOR'S WEBSITE
Creaselightning
www.creaselightning.co.uk
Mark Bolitho's website, featuring
his work

ORIGAMI SOCIETIES
Asociación Española de
Papiroflexia
www.pajarita.org
Spanish origami association

British Origami Society
www.britishorigami.info
One of the most established
origami societies, based in
the United Kingdom

Centro Diffusione Origami
www.origami-cdo.it
Italian origami society

Japan Origami Academic
Society
www.origami.gr.jp
Japanese origami society with
a good magazine on advanced
folding techniques

Mouvement Français des
Plieurs de Papier
www.mfpp-origami.fr
French origami association

Nippon Origami
Association
www.origami-noa.jp
Japanese origami association

Origami Australia
www.origami.org.au
Supporter of Australian nonprofit
origami groups and convention

Origami Deutschland
www.papierfalten.de
German origami association

OrigamiUSA
www.origamiusa.org
With headquarters in New York,
this society holds one of the
biggest origami conventions of
the year

OTHER ORGANIZATIONS
Color Tree Ltd
www.colortreelimited.co.uk
Good UK supplier of origami
paper and products

Escuela Museo Origami Zaragoza
www.emoz.es
Origami museum in
Zaragoza, Spain

John Gerard Paper Studios
www.gerard-paperworks.com
German paper makers with a
range of handmade papers

Origami Spirit
www.origamispirit.com
US-based origami blog and a
range of interesting projects

Origamido Studios
www.origamido.com
US-based paper-art studio that
also produces bespoke paper for
origami artists

Shepherds Falkiners Fine Papers
store.bookbinding.co.uk
UK-based supplier of fine papers

AUTHOR'S ACKNOWLEDGMENTS
Thanks to Marion, John, Simon, Beth, Luke, Alex, Talia, Nick, Jen, Annabelle, Jack, Olly, and Graham,
and to my friends and family for their support in my origami journey.